D0778781

ISOTOPE SHIFTS
IN ATOMIC SPECTRA

PHYSICS OF ATOMS AND MOLECULES

Series Editors

P. G. Burke, *The Queen's University of Belfast, Northern Ireland*
H. Kleinpoppen, *Atomic Physics Laboratory, University of Stirling, Scotland*

Editorial Advisory Board

R. B. Bernstein *(New York, U.S.A.)*
J. C. Cohen-Tannoudji *(Paris, France)*
R. W. Crompton *(Canberra, Australia)*
J. N. Dodd *(Dunedin, New Zealand)*
G. F. Drukarev *(Leningrad, U.S.S.R.)*
W. Hanle *(Giessen, Germany)*

C. J. Joachain *(Brussels, Belgium)*
W. E. Lamb, Jr. *(Tucson, U.S.A.)*
P.-O. Löwdin *(Gainesville, U.S.A.)*
H. O. Lutz *(Bielefeld, Germany)*
M. R. C. McDowell *(London, U.K.)*
K. Takayanagi *(Tokyo, Japan)*

A Continuation Order Plan is available for this series. A continuation order will bring delivery of each new volume immediately upon publication. Volumes are billed only upon actual shipment. For further information please contact the publisher.

ISOTOPE SHIFTS IN ATOMIC SPECTRA

W. H. KING

University of Newcastle-upon-Tyne
Newcastle-upon-Tyne, United Kingdom

WITHDRAWN
UTSA Libraries

PLENUM PRESS • NEW YORK AND LONDON

Library of Congress Cataloging in Publication Data

King, W. H., 1935–
  Isotope shifts in atomic spectra.

  (Physics of atoms and molecules)
  Bibliography: p.
  Includes index.
  1. Atomic spectra. 2. Isotope shift. I. Title. II. Series.
QC454.A8K56 1984   539.7   84-2142
ISBN 0-306-41562-3

© 1984 Plenum Press, New York
A Division of Plenum Publishing Corporation
233 Spring Street, New York, N.Y. 10013

All rights reserved

No part of this book may be reproduced, stored in a retrieval system, or transmitted
in any form or by any means, electronic, mechanical, photocopying, micrifilming,
recording, or otherwise, without written permission from the Publisher

Printed in the United States of America

LIBRARY
The University of Texas
At San Antonio

PREFACE

Atomic and nuclear physics are two flourishing but distinct branches of physics; the subject of isotope shifts in atomic spectra is one of the few that links these two branches. It is a subject that has been studied for well over fifty years, but interest in the subject, far from flagging, has been stimulated in recent years. Fast computers have enabled theoreticians to evaluate the properties of many-electron atoms, and laser spectroscopy has made it possible to measure isotope shifts in the previously unmeasurable areas of very rare isotopes, short-lived radioactive isotopes, weak transitions, and transitions involving high-lying atomic levels. Isotope shifts can now be measured with greater accuracy than before in both optical transitions and x-ray transitions of muonic atoms; this improved accuracy is revealing new facets of the subject.

I am very grateful to Dr. H. G. Kuhn, F.R.S., for having introduced me to the subject in the 1950s, and for supervising my efforts to measure isotope shifts in the spectrum of ruthenium. I thus approach the subject as an experimental atomic spectroscopist. This bias is obviously apparent in my use of the spectroscopist's notation of lower–upper for a transition, rather than the nuclear physicist's upper–lower. My reasons are given in Section 1.3 and I hope that nuclear physicists will forgive me for using this notation even for muonic x-ray transitions. My bias will not, I trust, be so obviously apparent in the weight given to different parts of the book. Following the advice of the King of Hearts, I have tried to begin at the beginning for the benefit of graduate research students starting to work in the subject and, although I did not go on till I came to the end and then stop, I have tried to give enough references for readers to get to the end if they so desire. The subject is still growing vigorously; at one time, when papers on isotope shifts in barium were published almost as quickly as I could read them, I felt the force of the Red Queen's remark that it takes all the running one can do just to keep in the same place.

I have tried to make a thorough review of all work published up to the end of 1982 and attempted to include all important and relevant papers. There are some references to papers published in 1983, but the review is not complete up to any particular date in 1983. No new experimental results (which will not be published elsewhere) are reported in the book, but some of the information derived from isotope shift work (particularly in Chapter 9) is new. For several elements I have reinterpreted one piece of published work in the light of another, or analyzed the published results in a new way.

I wish to record my thanks to Mrs. Susan Banks for converting my holograph into a typescript, and to Mrs. Dorothy Cooper for the careful preparation of all the figures.

CONTENTS

9. ISOTOPE SHIFTS AND OTHER RELEVANT WORK: AN ELEMENT BY ELEMENT REVIEW

INTRODUCTION

1.1. WHAT IS AN ISOTOPE SHIFT?

An atomic spectral line is characteristic of the element producing the spectrum. However, the energy of a spectral line does depend slightly on the isotope. Energy differences between isotopes are called isotope shifts. They occur in both atomic and molecular spectra but this book is concerned only with the former. If an atom has an upper level of energy E' and a lower level of energy E'', then a radiative transition between these levels will involve a photon of energy

$$\Delta E = E' - E'' = h\nu \tag{1.1}$$

and frequency ν, where h is the Planck constant.

In general, the energy of the transition is different for atoms of different nuclear mass but the same nuclear and electronic charge, i.e., the energy of the transition is different for different isotopes. If two isotopes are labeled H for heavy and L for light

$$E'_H - E''_H = h\nu_H \quad \text{and} \quad E'_L - E''_L = h\nu_L \tag{1.2}$$

and the difference $h\nu_H - h\nu_L = h\delta\nu$ is the isotope shift.

$$IS = (E'_H - E''_H) - (E'_L - E''_L) \tag{1.3}$$

In practice it is slightly preferable to work with term values rather than energies as explained in Section 1.3.1.

1.2. WHY IS THERE AN ISOTOPE SHIFT?

Each atomic level is described by an eigenfunction with eigenvalues of angular momentum and energy. The angular momentum of a level has a definite fixed value, so if the mass of the atomic nucleus is changed, by substituting one isotope for another, the energy of the level will have to change so that the angular momentum can remain the same. This change in energy is called a mass shift (MS) and is one major part of the isotope shift.

The other major part arises because the energy of an atomic level depends upon the size and shape of the electric charge distribution of the nucleus. Different isotopes have the same number of protons, but they do not have the same distribution of protons in space; the charge distribution is affected by the number of neutrons in the nucleus. The nuclear charge distribution gives an electric field which determines the energy of the atomic electrons. A change in the energy of a level arising from a difference in this field between isotopes is called a field shift (*FS*). To a good approximation the total isotope shift is the sum of these two effects

$$IS = MS + FS \tag{1.4}$$

The field shift is often referred to as the volume shift, but "field" is preferable since changes in shape as well as size of the nuclear charge distribution can lead to isotope shifts.

1.3. SIGN AND OTHER CONVENTIONS

1.3.1. SIGN CONVENTION FOR ISOTOPE SHIFTS

The custom, which will be followed in this book, is to describe an isotope shift in a transition as positive when the frequency of the line of the heavier isotope is greater. Thus

$$IS = h\nu_H - h\nu_L \tag{1.5}$$

The energy of a transition is the difference between the energies of two levels. These energies can be thought of as energies above the ground level or as term values, that is, binding energies below a relevant ionization limit. Using the term values T'' and T', of the lower and upper levels, respectively,

$$h\nu = T'' - T' \tag{1.6}$$

and

$$IS = (T''_H - T'_H) - (T''_L - T'_L). \tag{1.7}$$

The use of term values rather than energies has the slight advantage that the "lower-upper" convention is that almost invariably used by atomic spectroscopists to describe a transition, pace the International Union of Pure and Applied Physics, who would have them do the opposite. It is obvious that Eq. (1.7) can be rearranged to give isotope shifts in levels

$$IS = (T''_H - T''_L) - (T'_H - T'_L) \tag{1.8}$$

But T_H and T_L are measured from different zero points, the ionization limit of the heavy and light isotopes respectively. In general, there is no reason to consider the isotope shift to be zero at the ionization limit. This shift is zero only if the limit is the completely ionized limit where all Z electrons, Z being the atomic number, are at rest at infinity. In this state the isotopes differ only in their nuclear energies and these are not relevant to the subject of isotope shifts. So if T is taken as the term value from the completely ionized limit, then T_H and T_L are measured from a common zero point, and the shifts in the levels are

$$T_H'' - T_L'' = IS'' \quad \text{and} \quad T_H' - T_L' = IS' \tag{1.9}$$

Just as an isotope shift in a transition can be split into a mass part and a field part, so can the shift in a level. For example

$$IS' = MS' + FS' \tag{1.10}$$

On this sign convention, with term values measured downward from an ionization limit, the shift of a level is positive if the heavier isotope lies deeper, i.e., has a larger term value. It is another slight advantage of this convention that the mass shifts in transitions and levels of single electron atoms and ions (the simplest isotope shift of all—see Section 3.1) are positive. The field shift of a level is negative, except for those cases where the nuclear charge occupies a smaller effective volume in the heavier isotope.

1.3.2. OTHER CONVENTIONS

As in Section 1.1, a capital delta (Δ) will be used for changes in a transition between an upper and a lower level, whereas a small delta (δ) will be used for changes between isotopes.

The spectra of atoms at various stages of ionization will be labeled by Roman numerals. For example, "Ce IV" will be used to mean the fourth spectrum of cerium, that is to say the spectrum of Ce^{+++}. This number is given the symbol Z_a in formulas.

1.3.3. ERRORS AND UNCERTAINTIES

The errors and uncertainties given are either those in the paper from which they are taken, or the standard deviation if the paper makes it clear that the quoted uncertainties are not standard deviations. The uncertainties in the last quoted digits are given in brackets.

$$7480.4(7.0) \quad \text{means} \quad \pm 7.0$$

$$7480.4(7) \quad \text{means} \quad \pm 0.7$$

$$6.5371(16) \times 10^{-7} \quad \text{means} \quad \pm 1.6 \times 10^{-10}$$

1.4. UNITS

The energies of isotope shifts in optical transitions are small; they are of the same order of magnitude as magnetic hyperfine splittings. A large optical isotope shift would be only a few μeV, so isotope shifts are usually quoted as wave numbers σ (reciprocal wavelengths), or frequencies ν, rather than as energies

$$E = h\nu = hc\sigma \qquad (1.11)$$

where c is the speed of light in a vacuum. Wave numbers are usually given by spectroscopists in cm^{-1} and occasionally called kaysers. Isotope shifts in these units are invariably less than unity and have often been quoted in milli-kaysers, abbreviated to mK. The millikayser has not received the blessing of inclusion within the International System of Units, and isotope shifts are now more often given as a frequency than as a wave number. A typical large optical isotope shift might have the following values in the various units:

$$100 \text{ mK} = 0.1 \text{ cm}^{-1} = 10 \text{ m}^{-1} \equiv 3000 \text{ MHz} \equiv 12.3 \text{ } \mu\text{eV} \qquad (1.12)$$

In this book, most optical isotope shifts are given in terms of frequency, but a few extremely large shifts are given in units of cm^{-1}.

Isotope shifts are present in x-ray transitions for just the same reasons as they are in optical transitions. In these transitions, which involve inner electrons, the interactions with the nucleus are stronger and so the shifts are larger. X-ray isotope shifts are usually given in eV, the shifts being of the order of a fraction of an eV. An even larger interaction with the nucleus is obtained when the orbiting particle is a muon rather than an electron. In this case, also, the transition energy is in the x-ray part of the spectrum and the isotope shifts are usually given in eV, the shifts being of the order of a keV.

The relationships between various ways of expressing energies of photons are given in Appendix 1.

A SHORT HISTORY OF THE SUBJECT

2.1. HOW IT MIGHT HAVE STARTED

Isotope shifts in optical spectra are small—they are often considered as hyperfine structure. A typical size of an isotope shift is 3 GHz or so. For visible light of wavelength 500 nm or 5000 Å (Angstrom) this is 3×10^9 Hz in 6×10^{14} Hz, or in terms of wavelength, a typical shift is 2.5 pm or 0.025 Å. Such small shifts could not be observed before the advent of interferometric methods into spectroscopy. The first person to use such a method was Michelson, who with the aid of "the interferometer" was, in about 1890, deducing structures in spectral lines down to the 0.01 Å level. He was looking for a spectral line without structure whose wavelength could be used as an absolute standard of length. He proposed the cadmium red line for this (Michelson, 1893). It is probable that some of the structure in some of the lines which Michelson observed and rejected in his search for a length standard were due to shifts between isotopes. However, in view of the then current nature of atomic theory, we can hardly blame Michelson for his failure to discover isotopes via their isotope shifts.

2.2. ISOTOPE NONSHIFTS

Positive rays were discovered in 1886 by Goldstein and were used in 1913 by J. J. Thomson to show that isotopes occur among ordinary nonradioactive elements. Their existence among the radioactive elements had already been deduced from various pieces of evidence including spectroscopic results. The word "isotopic" had previously been used in the context of organic chemistry in 1904, but it was coined by Soddy (1913) with its current meaning. It is derived from the Greek *isos* (equal) and *topos* (place), the place being the place in the periodic table.

As the various naturally occurring radioactive substances were isolated and analyzed, it was realized that the same element was present in different places in different radioactive series. For instance, it was found that the so-called ionium and uranium X_1 were both chemically identical with thorium. Spectroscopy helped in these studies because identical spectra were evidence of chemical identity. For example, Russell and Rossi (1912) compared the spectra of ionium with thorium, and found them to be identical except for about five impurity lines. In that their conclusions were based on the identity of spectra of different isotopes, we could say that they were studying isotope nonshifts. Since the dispersion of their grating spectrograph was 2.6 Å per mm they had no chance of measuring an actual isotope shift.

2.3. EARLY WORK AND DEVELOPMENTS UP TO 1950

The advent of Bohr's theory of the structure of the atom (Bohr, 1913) suggested that the spectra of different isotopes should not be identical but should be mass dependent. Attempts were soon made to measure isotope shifts. The first attempt, using ordinary lead and lead of radioactive origin (Merton, 1915), failed to detect an isotope shift, even though a Fabry and Pérot etalon was used so that a wavelength difference of as little as 3 mÅ could have been detected.

The first successful measurement of an isotope shift was made by Aronberg (1918) who found that in the line of wavelength 4058 Å of the lead spectrum, the wavelength of radio-lead was longer than that of ordinary lead by 4.3 mÅ. It was soon realized that such large shifts were unlikely to be due to the difference in mass between isotopes (Ehrenfest, 1922), and Bohr (1922) suggested the possibility that the shifts were due to a difference in the internal nuclear structure between the isotopes, giving a slight difference in the field of force surrounding the nuclei.

This was the first suggestion of a field shift, and Bohr also pointed out that the effect would be largest in the case of electrons in s configurations. These are the only electrons with a large probability, $\psi^2(0)$, of being at the nucleus. The first quantitative formulations of the field shift were made by Rosenthal and Breit (1932) and Racah (1932), following the first calculation of the mass shifts in atoms with more than one electron by Hughes and Eckart (1930). By this time, isotope shifts had been measured in a number of heavy elements and it was realized that experimental results and calculated values from theory were not always in good agreement. Breit (1932) considered the measured shifts in spectra of mercury, thallium, and lead. The size of the field shift depends on the change, during the transition, in the probability of there being electrons (s electrons, in fact, the probability being negligible for other electrons) at the nucleus. Breit pointed out that the change in $|\psi(0)|^2$ was better obtained from magnetic hyperfine structure measurements than from *ab initio* calculations. He also pointed out that even in a "transition" such as $6p^2 - 6p8p$ in the spectrum of lead, where no change in $|\psi(0)|^2$ was to be expected because there was no change in the number of s electrons, there could be a change in $|\psi(0)|^2$ (of the s electrons) because the s electrons were screened to different extents by 6p and 8p electrons, respectively. This paper is also an early example of the technique of using measured isotope shifts to deduce the configurations involved in transitions which had not been analyzed.

A transition with a very large field shift must be one in which the number of s electrons changes.

Isotope shifts continued to be measured during the 1930s. As enriched isotopes were not then available, it was necessary to have narrow spectral lines and yet be able to study lines in which the isotope shift was large, lines which were often weak. The light source that was most frequently used to achieve this was the Schüler tube, or the hollow cathode lamp as it came to be known. This gave a good compromise of reasonably narrow and reasonably strong lines. The spectra were almost invariably observed in high resolution by photographing the fringes produced by a Fabry and Pérot etalon. Two good examples of the work of this period are Schüler and Korsching (1938) on ytterbium, and Kopfermann and Krüger (1937) on argon.

Another method of obtaining narrow lines suitable for isotope shift measurements is to use an atomic beam. This technique was applied to isotope shift work by Jackson and Kuhn. Although giving very narrow lines, the lack of intensity meant that the method was applicable only to resonance lines. A good example of their work is their measurement of isotope shifts in potassium (Jackson and Kuhn, 1938). It was at about this time that Vinti (1939) showed how better values for the mass shifts in the spectra of many-electron atoms could be calculated. His method did not come into its own until much later, when fast computers made it possible to work with Hartree–Fock and other more realistic wave functions.

The earliest work on isotope shifts had been mainly concerned with obtaining a better understanding of the electronic side of the electron–nuclear interaction. As Breit (1932) wrote, "The values of nuclear radii and their differences have a significance only so far as order of magnitude is concerned, on account of uncertainties in the values of $\psi^2(0)$". As the electronic side became better understood, the interest changed to the determination of the changes in nuclear charge distributions between isotopes. The analysis of Broch (1945) contributed to this better understanding.

Brix and Kopfermann (1949) pointed out that if a nucleus was deformed from a spherical shape, but had no spin, the charge distribution of the deformed nucleus must be averaged over all directions in space, as with no nuclear spin there is no preferred direction relative to the electrons for the nuclear deformation. The deformation has the same effect as an increase in nuclear volume, and the large field shifts between ^{152}Sm and lighter isotopes were attributed to a large deformation in ^{152}Sm. Isotope shifts were thus able to give information on nuclear shape as well as size. We can conclude the period up to 1950 by mentioning the paper of Crawford and Schawlow (1949). They put the subject onto an altogether more quantitative footing, by giving careful consideration to the approximations used in the perturbation method of the theory of field shifts, and by making quantitative estimates of the amount of screening of one electron by another. They were not completely successful, but they demonstrated the way the subject should, and indeed did, develop in the succeeding few years.

The experimental data available at about this time were collected together by Brix and Kopfermann (1952).

2.4. THE STORY SINCE 1950

By 1950 the subject of isotope shifts in atomic spectra was deemed to have come of age; a review article on the subject was published in *Reports on Progress in Physics* (Foster 1951). The subject developed rapidly because of improvements in experimental techniques. The availability of highly enriched isotopes meant that isotope shifts much smaller than the Doppler width could be measured with the aid of a reference line technique, and shifts involving rare as well as common isotopes could be measured. Detection of the spectra by a photomultiplier rather than a photographic emulsion enabled intensities and shapes of spectral lines to be measured more accurately. With the development of suitable computer programs it was possible to correct for the perturbations of impurity isotopes and hyperfine structure (hfs). These developments reduced the uncertainty in an isotope measurement from a few mK (to use the unit of the day) down to a fraction of a mK. Good examples of work done with these techniques are in samarium and neodymium spectra (Hansen *et al.*, 1967), and in the spectrum of tin (Silver and Stacey, 1973).

At the same time as what might be called the standard techniques were being steadily improved in accuracy, their range of application was widened to include radioactive isotopes. Work on the spectra of very heavy elements included the first measurement of isotope shifts involving radioactive isotopes. Such measurements were coincidental to the main thrust of the work; the first work on radioactive nuclides, in which the measurement of isotope shifts was a major consideration, was carried out in the spectrum of thallium (Hull and Stroke, 1961) and of mercury (Kleiman and Davis, 1963). The inclusion of radioactive isotopes allowed isotope shifts to be measured over long series of isotopes and to be compared for nuclides with the same number of neutrons but different numbers of protons. Such data were very useful in revealing some of the features of nuclear structure.

The emphasis in most isotope shift work carried out in the 1950s and 1960s was on the determination of changes in the size of the nuclear charge distribution between isotopes. The development of the theory of the subject concentrated mainly on answering the question as to exactly what property of the nuclear charge distribution it was that was being measured. Significant contributions were made by various people including Babushkin (1963), Bodmer (1959), and Fradkin (1962). The conclusion reached was that, to a good approximation, the field shift of a term was a product of an electronic factor and a nuclear factor, the nuclear factor being proportional to $\delta\langle r^2 \rangle$, the change between isotopes in the mean square radius of the nuclear charge distribution.

The main difficulty in determining $\delta\langle r^2 \rangle$ was the uncertainty in the knowledge of the electron screening, and in particular how it changed in a transition from one level to another. A screening factor β was often introduced to allow for this; it was hoped that β was unity, but it was feared that it might differ from unity by 20%. The way to reduce the uncertainty in β is to study simple transitions, and an extreme example of this approach is to measure isotope shifts in x-ray transitions. This was first done with a successful positive result, at the California Institute of Technology, by observing the K x-ray isotope shift between ^{238}U and ^{233}U (Brockmeier et al., 1965). Many such measurements have since been made there, and some at other places, but the technique has never proved quite as useful as it promised to be. The snag is that it is difficult to measure shifts in x-ray transitions with great accuracy, and the advent of laser spectroscopy in the optical part of the spectrum has made this lack of accuracy in x-ray transitions more apparent. It was concluded, however, that, in principle, x-ray isotope shifts can be interpreted without the problems associated with the interpretation of isotope shifts in optical spectra (Seltzer 1969).

This technique has also been somewhat overshadowed by the study of isotope shifts in muonic atoms. Here also, the transition energy is in the x-ray part of the spectrum, but the muon is so close to the nucleus (compared with an atomic electron) that the field shifts are much larger. There is no problem in allowing for the mass shift, as it is just the normal mass shift of a one-electron (muon actually) atom. It is indeed possible to determine the nuclear size for one particular isotope since the energy of such a simple system can be calculated from theory for the case of a hypothetical infinitely heavy point nucleus. Early work on muonic spectra was limited to the commonest isotope of an element because the isotope shifts could not be resolved. The first measurements of shifts between isotopes in muonic spectra were made in oxygen, calcium, tin, and gadolinium by Cohen et al. (1966) at Columbia University, New York.

Isotope shifts in x-ray spectra from either electronic or muonic atoms can easily be split up into mass shifts and field shifts with reasonable precision. This was not possible in the case of shifts in optical spectra, where the best that could be done was to assume that mass shifts were small relative to the field shifts for all but the light elements. This assumption had to be dropped when it was shown that even in the case of an element as heavy as samarium, some of the transitions had large mass shifts (King, 1963). Unfortunately there was, at that time, no way of knowing in which transitions the mass shifts were large. This became known only after Bauche (1966; 1969) had calculated some values of the mass shift using Hartree–Fock wave functions.

At this time, considerable light was thrown on the other main problem in the interpretation of isotope shifts in optical spectra, the electronic screening factor. Just as the advent of powerful computers and suitable programs had allowed Bauche to calculate mass shifts, so it allowed Wilson (1968) to calculate the screening effects of electrons in heavy atoms. He also used Hartree–Fock calculations and was able to interpret much isotope shift data in the spectra of platinum, mercury, and thallium. This work revealed some hitherto unsuspected screening effects.

This better understanding of the theory of optical isotope shifts was applied to experimental results in the 1970s. The accuracy of these results was also greatly improved by the use of laser-spectroscopic techniques of measurement. The first isotope shift to be measured with the help of a laser was in the spectrum of xenon (Brochard and Vetter, 1966). The accuracy of measurement was improved from about 0.1 mK, as in the best nonlaser work, to about 0.01 mK, i.e., from 3 MHz to 300 kHz. This great improvement in accuracy was expected to show up new effects and the expectation was gratified. However, laser techniques could be applied to only a few transitions in very few spectra before the tunable dye laser became available as a spectroscopic tool. These lasers have sufficient power and monochromaticity for various techniques to be employed to eliminate the problem of the Doppler width of spectral lines. Many such Doppler-free measurements of isotope shifts have been made with dye lasers since the first experiments on hydrogen (Hansch et al., 1974) and ytterbium (Broadhurst et al., 1974).

These accurate Doppler-free results have shown up various small effects in isotope shifts which could have been expected, such as what might be called the fine structure of isotope shifts (J-dependent results). They have also shown the unexpected result, in the case of samarium (Griffith et al., 1979), that the measured shift cannot be uniquely divided into the sum of a mass shift and a field shift (Palmer and Stacey, 1982).

Recent developments in the theory of isotope shifts have been concentrated into making (more or less pseudo-) relativistic calculations of mass and field shifts in heavy elements. Recent progress in the theory of the electronic side of isotope shifts is well reviewed by Bauche and Champeau (1976). The very detailed method of approach to the case of one-electron systems (Erickson, 1977) is now being applied to two-electron systems (Mårtensson, 1979) and will, it is hoped, gradually be applied to more complicated atoms. As the theory of the electronic side of isotope shifts has developed it has become of more and more interest to measure shifts in many lines of a spectrum. The technique of Fourier transform spectroscopy is ideally suited to this, and the technique has been used in measuring isotope shifts. The work of Gerstenkorn and Vergès (1975) on mercury is a good example.

In order to obtain information about nuclear charge distributions, it is necessary to study only one transition in which the electronic effects are well understood in any one element. The different types of transition (optical, electronic x-ray, and muonic x-ray) are all proving to have their uses in different circumstances, although the second is perhaps out of favor at the ,moment. There are, of course, other methods of determining nuclear charge distributions, electron scattering experiments being the most obvious, and the interpretation and reconciliation of the results from these different types of experiment has been a major preoccupation of theoretical nuclear physicists for many years. This subject was well reviewed by Barrett and Jackson (1977) in their monograph on nuclear sizes and structure.

As an example of what can be discovered about the sizes of the nuclear charge distributions of the isotopes of an element, the case of barium will be considered. The muonic x-ray spectra of five isotopes (^{138}Ba–^{134}Ba) have been studied (Shera et al., 1982). These have given various pieces of information about the nuclear charge distribution, including the mean square radii of the nuclear charge distribution which are shown in Fig. 2.1. This shows that the charge distribution of the nuclides gets larger as pairs of neutrons are added but that the odd nuclides have significantly smaller charge distributions than the neighboring even nuclides. The change in $\langle r^2 \rangle$ between ^{136}Ba and ^{134}Ba is $\delta\langle r^2 \rangle = 0.0063(53)$; the uncertainty is comparable with $\delta\langle r^2 \rangle$. Optical isotope shifts have also been measured between these isotopes, using laser-spectroscopic techniques. The shift between ^{136}Ba and ^{134}Ba is 15.3(18) MHz; the uncertainty is here much less than the shift. Unfortunately the optical isotope shifts cannot be converted into values of $\delta\langle r^2 \rangle$ with the same small uncertainty. This is because the isotope shift includes a mass shift, whose size is not known exactly, as well as a field shift. Also the conversion factor from field shift to $\delta\langle r^2 \rangle$ is not known with very high precision. The maximum amount of information about $\delta\langle r^2 \rangle$ is obtained by combining the muonic and optical isotope shift data. This shows that the mass shift in $\lambda 553.5$ nm of Ba I ($6s^2 {}^1S_0$–$6s6p\,{}^1P_1$), in which isotope shifts were measured by Baird et al. (1979) and Bekk et al. (1979), is 9 MHz. The method of obtaining this mass shift is explained in Section 6.2. A comparison of the field shifts in $\lambda 553.5$ nm with the $\delta\langle r^2 \rangle$ values obtained from muonic data, shows that the conversion factor from field shift to $\delta\langle r^2 \rangle$ is $3.2(1)\times10^{-4}$ fm^2/MHz as explained in Section 9.56. Bekk et al. (1979) studied the optical isotope shift of many unstable neutron deficient isotopes. These were produced as beams via compound nuclear reactions by charged particle irradiation of appropriate targets using the deuteron and α-particle beams of the Karlsruhe Isochronous Cyclotron. The isotope shift of radioactive ^{140}Ba was studied by Fischer et al. (1974a) who separated it chemically from fission products of uranium before measuring its spectrum. In Fig. 2.2 values of $\delta\langle r^2 \rangle$ relative to ^{138}Ba are shown for all

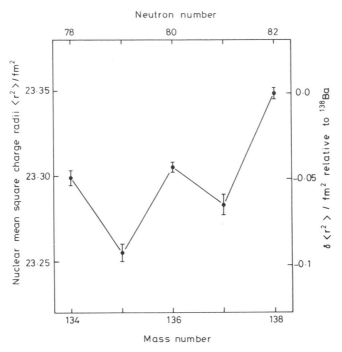

FIGURE 2.1. The mean square radii of the nuclear charge distribution of barium isotopes deduced from the energies of muonic x-ray transitions. Data from Shera *et al.* (1982).

these isotopes. For three isotopes, an excited nuclear state (isomer) was studied in addition to the ground state isotope. For further details of these results and their uncertainties see Section 9.56.

Fig. 2.2 reveals a number of features about the nuclear sizes of barium isotopes. The value of $\delta\langle r^2\rangle$ for ^{140,138}Ba is of a size that is typical for isotope shifts in this part of the periodic table. The very small values of $\delta\langle r^2\rangle$ between the lighter isotopes are believed to be so for two reasons. Firstly, the added neutrons are forming a neutron skin around the protons rather than mixing with them, and secondly, the deformation of the nuclides is decreasing (and so counteracting any increase in size between nuclides) as a magic number of neutrons is approached at ^{138}Ba where $N = 82$. The so-called "odd–even staggering effect" is very apparent in Fig. 2.2 for all the odd isotopes. The measurement of three isomer shifts shows that the amount of odd–even staggering is related to the neutron configuration; an effect that was first pointed out by Tomlinson and Stroke (1962) in the nuclides of mercury. In the case of ^{133}Ba and ^{135}Ba, the outer odd neutrons are in a lower angular momentum orbit when in the ground state than when in the isomeric state. The opposite is the case for the more strongly deformed ^{129}Ba nuclide.

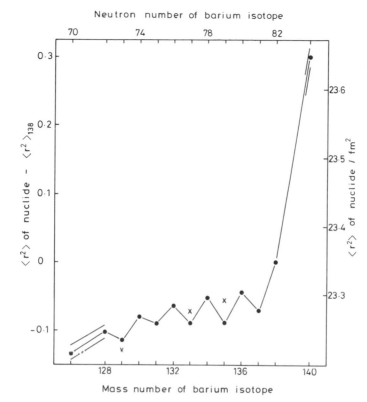

FIGURE 2.2. Mean square radii of the nuclear charge distribution of barium isotopes. Points labeled (●) are ground state nuclides; points labeled (×) are isomers. The scale on the right uses the value of $\langle r^2 \rangle$ for ^{138}Ba obtained by Shera et al. (1982) from transition energies in muonic atoms. The extra lines between ^{128}Ba and ^{126}Ba and near ^{140}Ba are explained in Section 9.56.

The example of barium shows the sort of information about nuclear structure that can be obtained with the help of isotope shift measurements. Not all elements have been studied so thoroughly, but several have, such as Ca, Cd, Sn, Ce, Nd, Sm, Os, Hg, Pb, and U. Isotope shifts have also been used as an aid to the elucidation of atomic electron structure from atomic spectra. At its crudest this involves simply saying that the largest (assumed field) shifts occur in the transitions with the largest effective change in the number of s electrons during the transition. Because of configuration interaction this change does not have to be integral. At its most elegant it involves a parametric analysis of the isotope shift results, as in, for example, the work of Grethen et al. (1980) in Pt I.

A few years ago the study of isotope shifts ceased to be an academic exercise aimed solely at the elucidation of the electronic and nuclear properties of atoms; it became a topic of importance in applied physics when it was realized that spectroscopic methods of isotopic enrichment could compete economically with the other, more traditional methods of enrichment. The principle is simply that, with sufficiently monochromatic light, one isotope of a natural mixture of isotopes can be preferentially excited by resonance absorption. The excited atoms are then picked off by selectively ionizing them or using some chemical process that discriminates between excited and ground state atoms. The advent of powerful tunable lasers turned the principle, which was recognized in the earliest days of isotope shifts (Hartley *et al.*, 1922), into a practical reality in the 1970s. So much so that isotope separation is today one of the most actively pursued areas in work on the application of lasers.

INTRODUCTION TO THE THEORY OF MASS SHIFTS IN OPTICAL SPECTRA

3.1. THE MASS SHIFT IN THE SPECTRA OF ONE-ELECTRON ATOMS AND IONS

This isotope shift is usually called the normal mass shift; it is also called the Bohr shift or the reduced mass effect. Consider an electron of mass m interacting with a point nucleus of mass M via their mutual electrostatic attraction. In classical mechanics the two particles can be treated as one particle with a reduced mass $mM/(M + m)$ interacting with an infinitely massive particle at rest. For a classical orbit of given shape and angular momentum, the term value or binding energy $T_{(M)}$ depends on the nuclear mass according to the relationship

$$T_{(M)} \propto \frac{M}{M + m} \tag{3.1}$$

The same result is obtained by solving the nonrelativistic Schrödinger equation, as is explained, for example, by Bethe and Salpeter (1977). It is sometimes useful to compare the term value with that which would arise from a hypothetical isotope of infinite nuclear mass, T_∞. The following results obviously follow from Eq. (3.1):

$$\frac{T_{(M)}}{T_\infty} = \frac{M}{M + m}, \qquad \frac{T_\infty - T_{(M)}}{T_\infty} = \frac{m}{M + m}, \qquad \frac{T_\infty - T_{(M)}}{T_{(M)}} = \frac{m}{M} \tag{3.2}$$

T_∞ is a useful theoretical concept, particular in more complicated atoms, but all that can be measured experimentally is the shift between a heavy isotope and a light isotope with nuclear masses M_H and M_L. Again it follows from Eq. (3.1) that

$$\frac{T_H - T_L}{T_H} = \frac{m(M_H - M_L)}{M_H(M_L + m)}, \qquad \frac{T_H - T_L}{T_L} = \frac{m(M_H - M_L)}{M_L(M_H + m)} \tag{3.3}$$

These results for term values apply equally well to transitions between terms, and so for the isotope shifts which can actually be measured experimentally in

15

a spectrum:

$$\frac{\nu_H - \nu_L}{\nu_H} = \frac{m(M_H - M_L)}{M_H(M_L + m)}, \qquad \frac{\nu_H - \nu_L}{\nu_L} = \frac{m(M_H - M_L)}{M_L(M_H + m)} \qquad (3.4)$$

This mass shift is called the normal mass shift, sometimes abbreviated to *NMS*. It is even less "normal" than the normal Zeeman effect since it completely describes the mass shift only in the spectra of one-electron atoms and ions. This theory has assumed the electron is in the electrostatic potential of a point nucleus but there are small additional contributions from relativistic and quantum electrodynamical effects; these can be ignored except in work of the highest precision.

The normal mass shift is easily calculated, and for hydrogen

$$\frac{T_{^2\mathrm{H}} - T_{^1\mathrm{H}}}{T_{i^1\mathrm{H}}} = \frac{\nu_{^2\mathrm{H}} - \nu_{^1\mathrm{H}}}{\nu_{^1\mathrm{H}}} = 2.721 \times 10^{-4} \qquad (3.5)$$

So in the Balmer α line, for example, the shift between $^2\mathrm{H}$ and $^1\mathrm{H}$ is 22.4 cm^{-1} or 671 GHz or 1.79 Å. The last figure is perhaps the most relevant; it shows that the shift is easily observable with a quite modest spectroscope. For the temperatures in normal laboratory light sources, the Doppler width of the $^1\mathrm{H}$ line is much less than this shift. Even so, the isotope effect is not normally observed because the natural abundance of deuterium is only 0.015%.

The existence of deuterium was first shown by the presence of its isotope-shifted spectrum beside that of $^1\mathrm{H}$ from an enriched source. Urey *et al.* (1932) enriched the rare isotope by a fractional distillation of liquid hydrogen. They obtained an enrichment of more than 0.1% deuterium and were able to measure the isotope shift in the Balmer α, β, γ, and δ lines to about ± 0.02 Å (about ± 7000 MHz). They obtained shifts that agreed with predictions from theory for an isotope of mass number 2. They used a photographic grating spectrograph in second order which gave a dispersion of 1.3 Å per mm.

A recent measurement by Wieman and Hänsch (1980) makes an interesting comparison. They found the shift between $^2\mathrm{H}$ and $^1\mathrm{H}$ in the Lyman α line to be 670,922(6) MHz. Doppler-free two-photon spectroscopy was used to achieve this precision, but even here the field shift is barely significant, being -6 MHz. More important corrections to the theory arise from relativistic effects, and this measurement was the first of sufficient accuracy to confirm the relativistic correction which must be applied to the nonrelativistic reduced mass theory outlined above. For Lyman α this correction is 9 MHz. With this and other terms correctly considered, it is found that the largest uncertainty in the theoretical value arises from the uncertainty in our knowledge of the electron to proton mass ratio. In the future, the best value for this ratio may be obtained from measurements of isotope shifts in hydrogen. Indeed the main reasons for this type of work is to improve our knowledge of the fundamental constants and to test quantum electrodynamics. It is really a different subject from the other isotope shifts considered in this book, and so it will be discussed no further. For more details see the references given by Wieman and Hänsch (1980), and in particular the review article on energy levels of one-electron atoms by Erickson (1977).

The lightest one-electron ion is He$^+$ and

$$\frac{T_{^4\mathrm{He}} - T_{^3\mathrm{He}}}{T_{^4\mathrm{He}}} = 4.485 \times 10^{-5} \tag{3.6}$$

So the normal mass shift of the ground level relative to the ionization limit is 19.69 cm^{-1} or 5.902×10^5 MHz.

Table 3.1 gives some representative values of normal mass shifts of one-electron ions for elements scattered through the Periodic Table. The normal mass shift in the ground level is about 13 cm^{-1} for a difference of one neutron between isotopes. This quantity is expected to be roughly constant throughout the periodic table as the fractional normal mass shift is proportional to M^{-2}, and the term value of the ground level is approximately proportional to Z^2. Z is the atomic number and Z/M is about 0.5 throughout the Periodic Table.

The field shift increases with Z, and for mercury the nonrelativistic field shift of the ground level is 1.7×10^7 MHz or 5.7×10^2 cm^{-1}. This is large in absolute terms, but is quite negligible relative to the term value of 7.7×10^8 cm^{-1} or 95 keV. This value is, of course, a calculated value from the theory, and not an experimental result. With $Z = 80$, and α the fine structure

TABLE 3.1
Normal Mass Shifts in Spectra of One-Electron Atoms and Ions

Heavy isotope	Light isotope	$\dfrac{T_H - T_L}{T_H \text{ or } T_L}$ [a]	Normal mass shift of ground level/cm^{-1}
$^2_1\mathrm{H}$	$^1_1\mathrm{H}$	2.721×10^{-4}	29.84
$^4_2\mathrm{He}$	$^3_2\mathrm{He}$	4.485×10^{-5}	19.69
$^7_3\mathrm{Li}$	$^6_3\mathrm{Li}$	1.302×10^{-5}	12.86
$^{22}_{10}\mathrm{Ne}$	$^{20}_{10}\mathrm{Ne}$	2.497×10^{-6}	27.43
$^{66}_{30}\mathrm{Zn}$	$^{64}_{30}\mathrm{Zn}$	2.601×10^{-7}	25.99
$^{202}_{80}\mathrm{Hg}$	$^{200}_{80}\mathrm{Hg}$	2.721×10^{-8}	21.06

[a] The isotopes used in the denominator are ^{1}H, ^{4}He, ^{7}Li, ^{20}Ne, ^{64}Zn, and ^{202}Hg.

constant, $Z\alpha = 80/137 = 0.58$, so relativistic and quantum electrodynamic effects are large. In fact, the nonrelativistic field shift is much smaller than the uncertainty in the quantum electrodynamic calculations of energies. This uncertainty is 2×10^5 cm^{-1} in the case of mercury.

As already mentioned, the isotope shift between deuterium and hydrogen is very large, but this section can be rounded off by pointing out that it pales into insignificance when compared with the isotope shift between hydrogen and positronium where, in the latter case, the nucleus is a positron. The isotope shift in the ground level is 54,810 cm^{-1}. (AUTHOR'S NOTE: This is a calculated value, and I can see no reason why anyone should bother to measure it accurately. Having written that, I suppose it will turn out to be a crucial test of QED theory.)

The theory of the one-electron atom is closely related to that of the muonic atom. Muonic atoms can be produced with many different nuclei and so the size of many nuclear charge distributions can be determined from the energy of the muonic x-ray transitions. This topic will be discussed in Chapter 5.

3.2. THE MASS SHIFT IN THE SPECTRA OF TWO-ELECTRON ATOMS AND IONS

If, in a two-electron atom, the electrons are uncorrelated, the energy of the atom is just the sum of the energies of two one-electron atoms and the term values are still given by Eq. (3.2)

$$T_\infty - T_{(M)} = \frac{m}{M + m} T_\infty \qquad (3.7)$$

The mass shift is just the normal mass shift. According to the virial theorem, the term value equals the mean kinetic energy of the atom since the atom is assumed to consist of particles (the finite size of the nucleus is ignored). For the atom with an infinitely heavy nucleus, its kinetic energy is the kinetic energy of its electrons, the kinetic energy of the nucleus being zero.

$$T_\infty - T_{(M)} = \frac{\langle p_1^2 + p_2^2 \rangle}{2(M + m)} \qquad (3.8)$$

$\mathbf{p}_1$ and $\mathbf{p}_2$ being the momenta of the electrons in the atom with an infinitely heavy nucleus. In general, the momentum of the nucleus $\mathbf{p}$, is given by

$$\mathbf{p} = \mathbf{p}_1 + \mathbf{p}_2 \qquad (3.9)$$

and

$$p^2 = p_1^2 + p_2^2 + 2\mathbf{p}_1 \cdot \mathbf{p}_2 \tag{3.10}$$

But in the absence of correlation between the electrons, the average value of $\mathbf{p}_1 \cdot \mathbf{p}_2$ is zero and

$$\langle p^2 \rangle = \langle p_1^2 + p_2^2 \rangle \tag{3.11}$$

So, in the absence of correlation, Eq. (3.8) becomes

$$T_\infty - T_{(M)} = \frac{\langle p^2 \rangle}{2(M + m)}$$

Experiment shows that there are, in general, small additions to this normal mass shift which arise because there *are* correlations between the electrons. This correlation is determined, not just by the energy of the level concerned, but by the specific properties of that level, so the extra mass shift is called the specific mass shift (*SMS*). It is small, compared with the normal mass shift of a level, and can be treated as a perturbation. Eq. (3.12) is taken to be still true, but $\langle p^2 \rangle$ is not given by Eq. (3.11) but is obtained from Eq. (3.10):

$$T_\infty - T_{(M)} = \frac{\langle p_1^2 + p_2^2 \rangle}{2(M + m)} + \frac{\langle \mathbf{p}_1 \cdot \mathbf{p}_2 \rangle}{M + m} \tag{3.13}$$

Even with correlation, it is still true for the atom with an infinitely heavy nucleus that its kinetic energy, which is the kinetic energy of its electrons, is equal to its term value, so

$$T_\infty - T_{(M)} = \frac{m}{M + m} T_\infty + \frac{\langle \mathbf{p}_1 \cdot \mathbf{p}_2 \rangle}{M + m} \tag{3.14}$$

Now $T_{(M)}$ is measurable whereas T_∞ is not, so Eq. (3.14) is more usefully rearranged as

$$T_\infty - T_{(M)} = \frac{m}{M} T_{(M)} + \frac{\langle \mathbf{p}_1 \cdot \mathbf{p}_2 \rangle}{M} \tag{3.15}$$

Note that the two terms are now only approximately equal to the normal and specific mass shifts respectively, but that both can be determined. The first term can be calculated to the accuracy with which M and $T_{(M)}$ are known. The second term can be calculated so long as the wave function is known; it is the expectation value of the specific mass shift operator $\mathbf{p}_1 \cdot \mathbf{p}_2 / M$.

It is $T_H - T_L$, not $T_\infty - T_{(M)}$, which is measured in an experiment. It follows from Eq. (3.15) that

$$T_H - T_L = \frac{m}{M_L} T_L + \frac{\langle \mathbf{p}_1 \cdot \mathbf{p}_2 \rangle}{M_L} - \frac{m}{M_H} T_H - \frac{\langle \mathbf{p}_1 \cdot \mathbf{p}_2 \rangle}{M_H} \qquad (3.16)$$

which can be rearranged to give

$$T_H - T_L = \frac{M_H - M_L}{M_H (M_L + m)} \left(m T_H + \langle \mathbf{p}_1 \cdot \mathbf{p}_2 \rangle \right) \qquad (3.17)$$

The first calculation of $\langle \mathbf{p}_1 \cdot \mathbf{p}_2 \rangle$ was done by Hughes and Eckart (1930) using linear combinations of products of hydrogenlike wave functions. They calculated the shift between ^{7}Li and ^{6}Li which had been measured by Schüler (1927) in Li II, the spectrum of Li$^+$. The strongest coupling between the electrons, which is all that was considered, arises from the exchange part of the electrostatic interaction. This coupling only arises if the orbital angular momentum quantum numbers of the two electrons differ by unity; see for example, Bethe and Salpeter (1977). Neglecting levels with both electrons excited, one electron is in the 1s configuration, and so the exchange shift occurs only in levels from the configuration 1snp. This shift is exactly opposite for singlet and triplet terms, being entirely an exchange effect. Hughes and Eckart (1930) calculated that between ^{7}Li and ^{6}Li the specific mass shift in the 1s2p^{3}P term was -0.85 cm^{-1}. Schüler (1927) had measured an isotope shift of 1.1 cm^{-1} in the lines 1s2s^{3}S$_1$–1s2p^{3}P which, since $\sigma = 18{,}227$ cm^{-1}, have a normal mass shift of 0.24 cm^{-1}. So the calculated shift of $0.24 + 0.85 = 1.09$ cm^{-1} was in good agreement with the experimental value. In Fig. 3.1 these shifts are shown relative to the completely ionized limit. This has been done to show that although the specific mass shift may be large compared with the normal mass shift in a transition, this is not true for a level. The specific mass shift in the 1s2p^{3}P term is only -0.85 cm^{-1} between ^{7}Li and ^{6}Li, whereas the normal mass shift from the completely ionized limit is 14.37 cm^{-1}.

If the two electrons were extremely coupled then either their momenta would cancel out so that there would be no nuclear motion, and hence no isotope shift, and $SMS = -NMS = -14.37$ cm^{-1} or in the other extreme, they would behave like one particle of mass $2m$, giving twice the normal mass shift, and so $SMS = NMS = 14.37$ cm^{-1}. Obviously the actual value of -0.85 cm^{-1} shows that the coupling between the electrons is weak and the mass shift of a level is approximately equal to the normal mass shift measured from the completely ionized limit.

The extreme case mentioned above, where the two electrons coalesce to give a large positive specific mass shift, could only occur for symmetric spatial

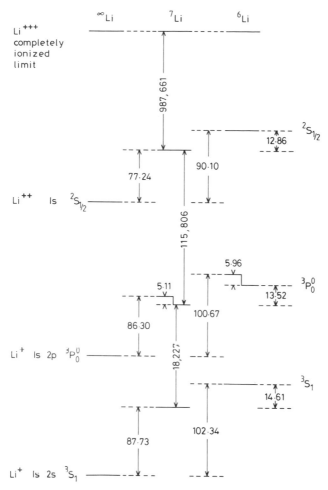

FIGURE 3.1. Calculated mass shifts in lithium (cm^{-1}). The shifts are shown (not to scale) relative to the completely ionized limit. The values are those calculated by Hughes and Eckart (1930) according to whom the only nonzero specific mass effect occurs in the $^3P_0^0$ level. In this level the normal mass effect is shown on the left and the normal plus specific mass effect is shown on the right. The shift in the transition $^3S_1 - ^3P_0^0$ is 1.09 cm^{-1} between ^{7}Li and ^{6}Li.

eigenfunctions, i.e., for antisymmetric spin functions (singlet terms). The specific mass shifts would thus be expected to be positive for singlets and negative for triplets, exactly what was predicted by Hughes and Eckart (1930), and in agreement with experimental results.

More recently the study of the spectra of two-electron atoms has concentrated on helium; but isotope shift measurements were not possible before the advent of enriched ^{3}He, because natural helium contains only 1 ppm of

^{3}He. It is unfortunately true, for helium as well as lithium, that one of the isotopes has an odd number of nucleons and so a nuclear magnetic moment. This leads to magnetic hyperfine structure, and so complicates the measurement of the isotope shift. It also leads to the problem of exactly what is meant by the isotope shift between the line of an even isotope and the hyperfine structure of an odd isotope. To achieve the precision of the work on lithium discussed so far, it is good enough to use the center of gravity of the hyperfine structure of the odd isotope, and this is what was done. In more precise work this is not good enough; the corrections for hyperfine structure that need to be made in very accurate isotope shift measurements will be discussed in Section 9.3.1. In reviewing more recent work on isotope shifts in spectra of two electron atoms, results in which these corrections have already been made will be used.

3.2.1. THE MASS SHIFT IN THE SPECTRUM OF NEUTRAL HELIUM (He I)

The calculation of the specific mass shift involves the calculation of $\langle \mathbf{p}_1 \cdot \mathbf{p}_2 \rangle$ and its value is very sensitive to details of the wavefunctions used. In order to make comparisons with recent experimental work (Freeman et al., 1980; de Clercq et al., 1981) it is necessary to use the most accurate wavefunctions available (Accad et al., 1971; Pekeris 1962; Schiff et al., 1965). The purpose of these calculations was, according to Pekeris, to do spectroscopy from scratch by solving the Schrödinger wave equation for two-electron atoms to an accuracy comparable with the experimental accuracy of the term values. Up to 560 terms in the expansion of the wave function have been used to obtain term values that agreed with experimental values to about 1 cm^{-1} or even 0.1 cm^{-1} in some cases. Wave functions were obtained for the case of an infinitely heavy nucleus and then the expectation values of the various relevant operators were computed. It was hence possible to calculate the correction in energy needed to go from an infinitely heavy nucleus to a nucleus of finite mass. Values were calculated of the mass polarization correction which is the specific mass shift from $^\infty$He to ^{4}He. These have been converted to the shifts between ^{4}He and ^{3}He in Table 3.2. Figures similar to these have been used by de Clercq et al. (1981) to make detailed comparisons with their experimental values of shifts in the $2\,^3$S–$n\,^3$S transitions. They used Doppler-free two-photon spectroscopy to obtain these results and also shifts in $2\,^3$S–$n\,^3$D transitions. No precise wave functions are available for ^{3}D levels, but using a multiconfigurational Hartree–Fock method suggests that the specific mass shift is less than 10 MHz in ^{3}D levels. If only the strongest electron coupling is considered, the exchange part of the electrostatic interaction, the specific mass shift in ^{3}S and ^{3}D levels, is zero. Freeman et al. (1980) used Doppler-free intermodulated fluorescence spectroscopy to measure the

TABLE 3.2
Calculated Specific Mass Shifts between ^{4}He and ^{3}He (GHz)

Principal quantum number of valence electron	Term[a]			
	^{1}S	^{3}S	^{1}P	^{3}P
1	46.937			
2	2.804	2.196	13.586	-19.053
3	0.78	0.56	4.293	-5.420
4	0.32	0.219	1.846	-2.229
5	0.21	0.107	1.0	-1.1
6	0.13	0.06		
7	0.09	0.03		

[a] Figures derived from calculations of Pekeris (1959), Pekeris (1962), and Accad *et al.* (1971).

isotope shift in the $2\,^3$P–$3\,^3$D transition. Both sets of results are summarized in Table 3.3.

The inadequacy of the Hughes–Eckart theory, which gives $-17,000$ MHz for the $2\,^3$P specific mass shift and zero for the others, is obvious. If we assume d to be zero, even the refined theory seems to be out by about -10 MHz. De Clercq *et al.* (1981) believe that the discrepancy can be accounted for by the relativistic corrections that arise in the theory of Stone (1961; 1963). They calculate that these relativistic corrections are -11 MHz for all of the transitions they studied except for $2\,^3$S–$4\,^3$S, where the correction is -10 MHz.

TABLE 3.3
Mass Shifts between ^{4}He and ^{3}He (MHz)[a]

Transition	Measured isotope shift (IS)	Isotope shift less normal mass shift ($IS-NMS$)	Calculated specific mass shift (SMS)	Residual shift ($IS-NMS-SMS$)
$2\,^3$S–$4\,^3$S	42,906(4)	1,971	1,977	-6
$2\,^3$S–$5\,^3$S	47,114(4)	2,078	2,089	-11
$2\,^3$S–$6\,^3$S	49,300(5)	2,128	2,137	-9
$2\,^3$S–$3\,^3$D	37,480(7)	2,188	$2,196-d_3$[b]	$-8+d_3$
$2\,^3$S–$4\,^3$D	44,661(7)	2,184	$2,196-d_4$	$-12+d_4$
$2\,^3$S–$5\,^3$D	47,984(8)	2,182	$2,196-d_5$	$-14+d_5$
$2\,^3$S–$6\,^3$D	49,794(8)	2,186	$2,196-d_6$	$-10+d_6$
$2\,^3$P–$3\,^3$D	3,805(12)	$-19,059$	$-19,053-d_3$	$-6+d_3$

[a] These figures have been derived from the work of de Clercq *et al.* (1981) and Freeman *et al.* (1980).
[b] d is the unknown specific mass shift in the relevant ^{3}D level.

The inadequacy of the Hughes–Eckart theory had already been known for many years, as it was shown by the experimental results of Bradley and Kuhn (1951). This work is particularly notable as it revealed how the isotope shifts varied as the ionization limit was approached. The ionization limit of helium is single electron He^+. This has only a normal mass shift, and not a specific mass shift, and so it was possible to determine experimentally the specific mass shifts in individual levels as well as in transitions between pairs of levels by extrapolating the experimental results to the ionization limit. Four different series were investigated. The $2\,^3S-n\,^3P$ series measurements have been superseded by the more recent work already mentioned. The $2\,^1P-n\,^1D$ series measurements were distorted by a perturbation of the ^1D levels by ^3D levels. In some cases the perturbation for ^3He differs from that for ^4He. The results for the remaining two series, $2\,^1P-n\,^1S$ and $2\,^1S-n\,^1P$ are plotted in Fig. 3.2. The agreement with the calculated values of Pekeris (1962) and Accad *et al.* (1971) is satisfactory, but not if the extrapolation made by Bradley and Kuhn (1951) to the limit $2\,^1S-\infty\,^1P$ is used. Their extrapolation is shown in Fig. 3.2 as a dashed line giving the specific mass shift of $2\,^1S$ as 4.0 GHz rather than the 2.8 GHz predicted by Accad *et al.* (1971).

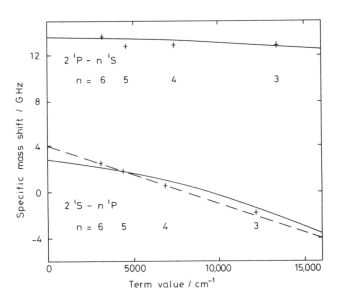

FIGURE 3.2. Specific mass shifts between ^4He and ^3He (GHz). The experimental values are the measured shifts of Bradley and Kuhn (1951) less the normal mass shifts. The full curves pass through the calculated values of Pekeris (1962) and Accad *et al.* (1971). The dashed line shows the linear extrapolation made by Bradley and Kuhn (1951). The specific mass shift of any level can be found since the specific mass shift of the ionization limit is zero, the limit being the single-electron ion He^+.

3.2.2. THE MASS SHIFT IN THE SPECTRUM OF Li^+ (Li II)

Since the pioneering work, already mentioned, of Schüler (1927) and Hughes and Eckart (1930), there have been very few measurements of isotope shifts in Li II (the second spectrum of lithium, i.e., the spectrum of Li^+). The interest has been much more in hyperfine structure, but the isotope shift in the $2s\,^3S\text{--}2p\,^3P$ transitions has recently been measured by Bayer *et al.* (1979). They used dye laser saturation spectroscopy in a low-energy Li^+ ion-beam and fluorescence light detection. The $2\,^3S_1\text{--}2\,^3P_0$ transition shows the isotope shift most clearly because the $2\,^3P_0$ state has no hfs splitting. They obtained a value for the isotope shift between ^7Li and ^6Li in this transition of 34,750 MHz, but a more recent determination has given 34,766(3) MHz.* The normal mass shift is 7,114 MHz and the specific mass shift is 27,635 MHz. This leaves a shift of 17(5) MHz, which might be explained by relativistic corrections, but is more likely to be a field shift. Assuming it is a field shift, it follows that the nuclear charge radius of ^6Li is greater than that of ^7Li, a result corroborated by electron scattering measurements. Neumann has calculated that $\delta\langle r^2\rangle = -1.7$ fm^2.

3.3. THE MASS SHIFT IN THE SPECTRA OF MULTIELECTRON ATOMS AND IONS

If, in a multielectron atom with an infinitely heavy nucleus, the ith electron has a momentum $\mathbf{p}_i$, the results of Section 3.2 for two electrons can be generalized by replacing $p_1^2 + p_2^2$ by $\Sigma_i p_i 2$ and $\mathbf{p}_1 \cdot \mathbf{p}_2$ by $\Sigma_{i>j}\mathbf{p}_i \cdot \mathbf{p}_j$. For example, the mass shift between an infinitely heavy isotope and an isotope of nuclear mass M is

$$T_\infty - T_{(M)} = \frac{m}{M} T_{(M)} + \frac{\langle \Sigma'_{i>j}\mathbf{p}_i \cdot \mathbf{p}_j\rangle}{M} \tag{3.18}$$

It follows from Eq. (3.13) that all mass shifts have the same dependence on the nuclear mass M; $T_\infty - T_{(M)}$ is inversely proportional to $M + m$. If three isotopes of nuclear mass M_H, M_I, and M_L have term values T_H, T_I, and T_L, then if the differences between the term values are mass shifts (including specific mass shifts), it follows from Eq. (3.13) that

$$\frac{T_H - T_I}{T_I - T_L} = \frac{(M_H - M_I)(M_L + m)}{(M_I - M_L)(M_H + m)} \tag{3.19}$$

This relation can be used to check if measured isotope shifts are entirely mass

---

* My thanks to Dr. R. Neumann for this information prior to publication.

shifts. In practice it is not always a sensitive test because field shifts sometimes obey Eq. (3.19), at least approximately.

The first successful calculation of the specific mass shift for an atom with more than two electrons was made by Vinti (1939) for Mg I (the first spectrum of magnesium, i.e., the spectrum of neutral magnesium).

Vinti calculated the first-order perturbation value of the specific mass shift from the specific mass shift operator

$$\frac{1}{M} \sum_{i > j} \mathbf{p}_i \cdot \mathbf{p}_j$$

He used wave functions which omit angular correlation and so the operator is nonzero only when the orbital angular momentum quantum numbers of the two electrons differ by unity. Russell–Saunders coupling was assumed, and spin–orbit forces were neglected so that the specific mass shifts were the same for all the J values within a multiplet. In Table 3.4, Vinti's calculated values are compared with the experimental values then available. The agreement was not very close but, as Vinti pointed out, "the explicit radial functions are probably not very accurate." He was perhaps 25 years ahead of his time, and the calculation of accurate specific mass shifts did not become feasible until the advent of fast computers. It can be seen that even Vinti's functions enabled him to predict the sign and relative magnitudes of the specific mass shifts in different transitions. There is no simpler way of getting these figures, as the electron coupling depends on fine details of the wave functions.

The method has been developed by Stone (1959) and especially by Bauche (1966); the developments are discussed in detail by Bauche and Champeau (1976). The Hartree–Fock method is the one most widely used for the determination of atomic wavefunctions suitable for the calculation of specific mass shifts. The advent of fast computers in the 1960s allied with suitable computer codes such as that of Fischer (1970) made it possible to calculate the relevant radial functions to a useful level of precision. The

TABLE 3.4
Specific Mass Shifts between ^{26}Mg and ^{24}Mg (GHz)[a]

Transition	Calculated specific mass shift	Measured isotope shift less normal mass shift
$3s^2\,^1S-3s3p\,^1P$	−0.28	−0.84
$3s^2\,^1S-3s3p\,^3P$	0.35	1.35
$3s3p\,^3P-3s3d\,^3D$	−1.18	−1.35
$3s3p\,^1P-3s3d\,^1D$	0.71	1.90

[a]The calculated values are those of Vinti (1939) and the experimental values are those which were available to him at the time.

TABLE 3.5

Specific Mass Shifts between ^{62}Ni and ^{60}Ni (GHz)[a]

Level	Calculated specific mass shift		Experimental estimate of specific mass shift relative to $3d^84s^2\,^3F$
	Absolute	Relative to $3d^84s^2\,^3F$	
$3d^94p\,^3P$	-701.62	-2.66	-2.2
$3d^84s4p\,^5D$	-699.23	-0.27	-0.5
$3d^{10}\,^1S$	-702.82	-3.86	-3.0
$3d^94s\,^3D$	-701.33	-2.37	-1.9
$3d^84s^2\,^3F$	-698.96	0.00	0.0

[a] Data obtained from Bauche and Champeau (1976).

specific mass shifts sometimes depend more on the electron configuration than on the values of orbital and spin angular momentum within one configuration. For example, transitions of the type 3d–4s have a large specific mass shift because of the change in the number of 3d electrons (Bauche and Crubellier, 1970); an experimental fact noted in the spectra of nickel (Schroeder and Mack, 1961) and copper (Wagner, 1955). The comparison of theoretical calculations with measured shifts for nickel is shown in Table 3.5. The approximate field shifts have been subtracted from the experimental results; they are only of the order of 0.2 GHz.

It can be seen that the agreement between theory and experiment is good. No such agreement can be obtained by simple "common sense" arguments. For instance, the contribution of the eight 3d electrons to the specific mass shift is about 100 GHz, so one might expect a difference of one 3d electron to give a specific mass shift of about 12 GHz. In fact, relaxation of the core and self-screening of the 3d electrons reduces it to about 2 GHz. Nickel has been dealt with first of all, because it demonstrates one of the successes of the Vinti method as used with Hartree–Fock wavefunctions by Bauche. In general, the method gives the right sign for a specific mass shift but tends to overestimate their magnitudes.

A rather similar program of theoretical research into specific mass shifts has been proceeding more or less independently in the U.S.S.R. A recent publication from this source which gives references to earlier work is that of Vizbaraite *et al.* (1979).

3.3.1. THE MASS SHIFT IN THE SPECTRUM OF NEUTRAL LITHIUM (LI I)

Using wave functions of the Hartree–Fock type, which are based on a central-field model and omit angular correlation, it is found that the only configuration of lithium which has a specific mass shift is $1s^2 np$, since no other configuration (ignoring double-electron excitation) has two electrons with angular momentum quantum numbers l, differing by unity. The experimental results of Hughes (1955) showed that this approximation is inadequate. For example, measurements made on the transitions $2\,^2S-2\,^2P$ and $2\,^2P-5\,^2D$ show that the specific mass shift in 2^2S-5^2D is not zero but 1.13(15) GHz between ^7Li and ^6Li. Some, but not all, of the results of Hughes (1955) have been superseded by more recent and accurate measurements. The currently known values of the specific mass shifts in lithium are collected in Table 3.6. The similarity of the shifts in the transitions from $2s\,^2S$ to $3d\,^3D$, $4d\,^3D$, and $5d\,^3D$ suggests, as was pointed out by Hughes (1955), that the specific mass shifts in the D terms are almost the same as the specific mass shift in the ionization limit $1s^2\,^1S_0$ of Li^+. The best estimate of this shift, which has not been measured experimentally, is probably that of Pekeris (1958) whose value for the mass polarization correction of 4.960 cm^{-1} leads to a calculated specific mass shift between ^7Li and ^6Li of 24.75 GHz for $1s^2\,^1S_0$ of Li^+.

Various attempts have been made to calculate the specific mass shifts, taking into account the angular correlation. The methods used for two-elec-

TABLE 3.6
Experimental Specific Mass Shifts between ^7Li and ^6Li (GHz)

Upper term	Isotope shift less normal mass shift[a]	
Li^+ $1s^2\,^1S$	$1.100(6)^f$	$-3.616(6)^f$
Li 5d ^2D		$-3.57(12)^b$
Li 5s ^2S		$-3.55(12)^b$
Li 4d ^2D	$1.101(6)^e$	
Li 4s ^2S	$0.994(6)^e$	
Li 3d ^2D	$1.105(4)^d$	
Li 3p ^2P	$2.4(4)^b$	
Li 3s ^2S		$-3.88(6)^b$
Li 2p ^2P	$4.716(5)^c$	
Li 2s ^2S		
	$2s\,^2S$	$2p^2P$
	Lower term	

[a] This is approximately the specific mass shift since the field shifts are probably less than 0.01 GHz (much less for transitions involving no change in the number of 2s electrons).
[b] Hughes (1955) using a hollow cathode and Fabry–Pérot interferometer.
[c] Mariella (1979) using laser-induced fluorescence of a lithium beam.
[d] Kowalski *et al.* (1978) using two-photon dye laser spectroscopy.
[e] Lorenzen and Niemax (1982) using two-photon dye laser spectroscopy.
[f] Assuming the shifts in the D terms are similar to and converging towards the shift in the ionization limit (see text).

TABLE 3.7
Specific Mass Shifts between ^{7}Li and ^{6}Li (GHz)

Term of Li I	Calculated specific mass shift[a]		Experimental value from Table 3.6
6d ^{2}D		−0.0001	
6p ^{2}P	−0.132		
6s ^{2}S	0.024		
5d ^{2}D		−0.0002	−0.05(12)
5p ^{2}P	−0.227		
5s ^{2}S	0.042		−0.07(12)
4d ^{2}D		−0.0006	−0.001(6)
4p ^{2}P	−0.442		
4s ^{2}S	0.088		0.106(6)
3d ^{2}D		−0.002	−0.005(6)
3p ^{2}P	−1.034		−1.3(4)
3s ^{2}S	0.230		0.26(6)
2p ^{2}P	−3.273 −4.607[b]		−3.616(6)
2s ^{2}S	0.962 0.994[b]		1.100(6)

[a] The shifts are relative to the ionization limit $1s^2$ ^{1}S of Li$^+$ and were calculated by Mårtensson and Salomonson (1982) except for those marked *b*.

[b] Prasad and Stewart (1966). According to Mårtensson and Salomonson, their figure of −4.607 for 2p^{2}P becomes −3.495 if 1s2p → dp and 1s2p → pd contributions are added. The inclusion of higher order terms would then bring this into good agreement with the experimental value of −3.616.

tron atoms (Section 3.2.1) have not yet been applied to multielectron atoms. Prasad and Stewart (1966) have used the configuration interaction approach, and Mårtensson and Salomonson (1982) have achieved great success using many-body perturbation theory. The latter treated the pair correlation by numerical solution of the "pair equation" (Mårtensson 1979). In any transition the change in the valence electron orbital produces a change in the core and the latter can make an important contribution to the specific mass shift, as already mentioned in the case of nickel. However, for lithium the core consists only of s electrons and so does not contribute to the specific mass shift. The specific mass shift of the *ns* terms is zero in first-order results and the calculated values in Table 3.7 of Mårtensson and Salomonson (1982) are second-order results. Although the agreement with experiment is good here, the second-order results for *n*p terms do not help at all; they are about one-tenth of the first-order values and worsen the agreement with experiment. In the case of the d terms the result is again entirely second order. The individual contributions for the 1s*n*d → p^2 and 1s*n*d → pf excitations are 50 to 100 times larger than the final result; they are nearly equal in magnitude but of opposite sign. Such cancellations mean that even higher order corrections may be important and much useful work remains to be done on the calculation of specific mass shifts in other light elements.

3.3.2. THE MASS SHIFT IN THE SPECTRA OF OTHER LIGHT ELEMENTS

By the term light elements is meant, in this context, those in which the field shift in a transition is much less than the mass shift; then the measured shift less the normal mass shift, sometimes called the residual shift, can be treated as a reasonable approximation to the specific mass shift. Sorting out the mass shifts from the field shifts is not trivial, and will be discussed in Chapter 6. However, in the case of the alkali metals, it is possible to make some estimates and these are given in Table 3.8. Obviously, the point at which the field shift becomes significant depends on the precision with which the isotope shifts are measured. In the case of sodium the field shift is well over 1% of the mass shift so let us, very arbitrarily, say that this is significant, and define the light elements, according to the above definition, to be from hydrogen through neon. No recent experimental work has been done on the isotope shifts in the spectra of any of these elements except neon. Some of the earliest laser-spectroscopy techniques were applied to neon and an attempt to explain the results was made by Keller (1973). He found that *ab initio* calculations of the specific mass shifts using the multiconfigurational Hartree–Fock method were quite disappointing. He was, however, able to make considerable progress by means of a parametric study. Even though the *ab initio* calculations are impossibly difficult to carry out, it is possible to say what parameters will be relevant in calculating the specific mass shifts between certain configurations or certain terms. The sizes of these parameters can be found if the experimental results are sufficiently precise and are available for a sufficiently large number of different transitions. Obviously, this number must be larger than the number of parameters, so the method can only be applied where the configuration mixing does not involve a large

TABLE 3.8
Approximate Isotope Shifts for Alkali Metal Spectra (GHz)

Alkali metal	Change in A	Transition	NMS	SMS	FS
$_1$H	1	$1s\,^2S-2p\,^2P$	671	0	-0.006
$_3$Li	1	$2s\,^2S-2p\,^2P$	5.82	4.72	0.005
$_{11}$Na	1	$3s\,^2S-3p\,^2P$	0.55	0.2	-0.01
$_{19}$K	2	$4s\,^2S-4p\,^2P$	0.27	-0.02	-0.02
$_{37}$Rb	2	$5s\,^2S-5p\,^2P$	0.06	<0.02	-0.1
$_{55}$Cs	2	$6s\,^2S-6p\,^2P$	0.02	<0.02	-0.3

number of different configurations. The parametric method is expertly dealt with by Bauche and Champeau (1976).

It was found that neon met the requirements for a parametric study. It was also apparent from the experimental results that the specific mass shift was J-dependent within a configuration. Eleven transitions of the $2p^5 3d - 2p^5 4p$ type were analyzed which had specific mass shifts ranging from 38 MHz to 97 MHz. With seven parameters for the $2p^5 4p$ configuration and two parameters for the $2p^5 3d$ configuration the experimental shifts could be predicted to about 0.3 MHz. The parameters included one for each Russell–Saunders term and two that involved the spin–orbit integral. In contrast, an *ab initio* computation of the shift between the centers of gravity of the configurations gave 93 MHz, in poor agreement with the empirical parametric result of 57 MHz. The variation of the specific mass shift from one Russell–Saunders term to another was attributed by Keller (1973) to the crossed second-order effects of the specific mass shift operator $\sum_{i>j} \mathbf{p}_i \cdot \mathbf{p}_j$ and the electrostatic energy G. The variation with J within one term was attributed to this, and also to relativistic and other corrections.

More recent measurements have been made on the rare isotope ^{21}Ne as well as on ^{20}Ne and ^{22}Ne. Biraben *et al.* (1975) used Doppler-free two-photon spectroscopy, and Julien *et al.* (1980) used the velocity-selective optical pumping method. Neither found any shifts that were inconsistent with the mass shift formula, Eq. (3.19).

Ab initio calculations have not been completely devoid of success. Biraben *et al.* (1976) measured isotope shifts between ^{22}Ne and ^{20}Ne using Doppler-free two-photon spectroscopy in the four transitions $2p^5(^2P_{3/2})3s[\frac{3}{2}]_2 - 2p^5(^2P_{1/2})4d$. Three of the transitions had the same shift, within experimental error, of 2781(3) MHz. The fourth transition, with upper level $2p^5(^2P_{1/2})4d[\frac{3}{2}]_1$, had a shift of 2765(3) MHz, significantly different from the others. The difference arising from the normal mass shift is less than 1 MHz, the experimental difference must be a specific mass shift effect. The unique transition is the only one with $J = 1$ for the upper level; it is the only one where $^1P_1$ is present in its intermediate-coupling expansion. Biraben *et al.* (1976) calculated *ab initio* the specific mass shift arising from $^1P_1$ in the upper level in the Hartree–Fock scheme and obtained the value of 18 MHz in good agreement with the experimental difference of 16(4) MHz. The normal mass shift of the transitions is 2525 MHz so there is a specific mass shift of 256 MHz present in the three transitions with the same isotope shift.

The work has been extended by Giacobino *et al.* (1979) to other upper levels, namely $2p^5(^2P_{3/2})4d$, $2p^5(^2P_{1/2})5s$, and $2p^5(^2P_{3/2})5s$. As before, they found that the shift was always different when $J = 1$ in the upper level. Whereas previously they had used the amount of ^1P present to calculate the difference of isotope shift when $J = 1$, they now used the measured isotope shifts to determine the amount of ^{1}P present in various levels. They obtained excellent agreement with theoretical predictions of $(\tau_1|^1\text{P})^2$, the square of the component of the wave function of the excited state $|\tau, J = 1\rangle$ on the ^{1}P state.

Unfortunately, they were not able to explain from theory the difference of the specific mass shift between different configurations; there is still useful work to be done in this field.

3.3.3. THE MASS SHIFT IN THE SPECTRA OF HEAVIER ELEMENTS

In heavier elements the situation is confused because there are field shifts present as well as mass shifts. The separation of mass shifts from field shifts will be discussed in Chapter 6 and the results obtained for different elements will be dealt with in Chapter 9. The pioneering work of Vinti (1939) on magnesium has already been mentioned. The large mass shifts in nickel have also been mentioned already. For heavier elements it was assumed that the mass shifts were negligible compared with the field shifts, until it was pointed out (King, 1963) that for samarium there are mass shifts in some optical transitions that are several times as large as the normal mass shift. Even for an element as heavy as samarium, the specific mass shift is sometimes not negligible compared with the field shift. It was shown by Bauche (1969) that these large specific mass shifts arise when there is a change in the number of 4f electrons.

3.3.4. COMPARISON BETWEEN ELEMENTS

It follows from Eq. (3.13), generalized to the multielectron atom, that $(T_\infty - T_{(M)})(M + m)/m$ is independent of M, and so is suitable for making comparisons between different elements with very different values of M. The specific mass shift part of this quantity is usually called, following Vinti (1939), the k factor

$$k = \frac{1}{m}\left\langle \sum_{i > j} \mathbf{p}_i \cdot \mathbf{p}_j \right\rangle \qquad (3.20)$$

It follows from Eq. (3.19) that the specific mass shift between heavy and light isotopes is related to the k factor by

$$SMS_{H-L} = \frac{(M_H - M_L)mk}{(M_L + m)(M_H + m)} \qquad (3.21)$$

Some k factors are given in Table 3.9 in Hartree's atomic units (a.u.), one of which equals two Rydberg constants (fixed nucleus)

$$1 \text{ a.u.} = 6.57968 \times 10^{15} \text{ Hz} \qquad (3.22)$$

The calculated k factors of terms are mainly values calculated by Bauche; they are relative to the completely ionized limit. The measured k factors of transitions have been chosen because they are large for the elements concerned; most specific mass shifts are much smaller than these.

Through most of the Periodic Table, the specific mass shifts of terms are in the opposite direction to the normal mass shifts. That means $\langle \Sigma \mathbf{p}_i \cdot \mathbf{p}_j \rangle$ opposes $\langle \Sigma p_i^2 \rangle$, i.e., the electron momenta tend to cancel each other out because of correlation between them. The amount of correlation is not large, however. For example, in the case of magnesium, $k = -27.6$ a.u., whereas the equivalent to k for the normal mass shift, the term value relative to the completely ionized limit, is 200 a.u.; in the case of nickel, $k = -360$ a.u. is to be compared with about 1500 a.u. for the normal mass shift equivalent. The ratio of the k factor to the term value increases with M, but even for plutonium it is probably not twice the value for nickel, i.e., it is probably less

TABLE 3.9
Specific Mass Shift k Factors (a.u.)

Element	Term	k factor (calculated)	$Z^{2.4\,a}$	$k/Z^{2.4}$
$_2$He	$1s^2\,^1S$	0.156	5.278	0.030
$_3$Li	$2s\,^2S$	0.301	13.97	0.021
$_{10}$Ne	$2p^5 3s\,^3P$	-14.1	251.2	0.056
$_{12}$Mg	$3s^2\,^1S$	-27.6	289.1	0.071
$_{20}$Ca	$4s^2\,^1S$	-130	1,326	0.098
$_{28}$Ni	$3d^8 4s^2\,^3F$	-360	2,973	0.121
$_{42}$Mo	$4d^5 5s\,^7S$	-1156	7,867	0.147
$_{62}$Sm	$4f^6 6s^2\,^7F$	-3374	20,033	0.168
$_{94}$Pu	$5f^6 7s^2\,^7F$	-10454	54,388	0.192

Element	Transition	k factor (measured)
$_2$He	$1s2p\,^3P - 1s3d\,^3D$	0.063
$_3$Li	$2s\,^2S - 2p\,^2P$	0.055
$_{10}$Ne	$3s[\frac{3}{2}]_2 - 4d'[\frac{5}{2}]_3$	0.016
$_{12}$Mg	$3s^2\,^1S_0 - 3s3p\,^1P_1$	-0.038
$_{28}$Ni	$3d^8 4s^2 - 3d^9 4p$	1.2
$_{62}$Sm	$4f^6 6s^2 - 4f^5 5d6s^2$	-2.00

[a] This is roughly proportional to the term value relative to the completely ionized limit.

than 50%. For a heavy element, like Pu, relativistic effects are very important, and the nonrelativistic k factor of Table 3.9 is probably only a poor approximation to the truth anyway. Relativistic effects are considered in the next section.

3.4. RELATIVISTIC EFFECTS

The situation is unsatisfactory from both an experimentalist's and a theoretician's point of view. The effects will be largest in the heavy elements, just when it is most difficult to disentangle the mass shifts from the field shifts. Also, in the heavy elements, it is difficult to compare experimental results with predictions from theory because there is often too much configuration interaction present for theoreticians to make realistic calculations. The direct determination of relativistic mass shifts is not yet possible and the best that theoreticians can do is to calculate relativistic corrections to a nonrelativistic result.

Relativistic effects were considered by Stone (1961; 1963) and his ideas have since been developed (Bauche and Champeau, 1976). The application of these to helium by de Clercq et al. (1981) has already been mentioned in Section 3.2.1. Relativistic effects lead to a J dependence of the isotope shift, and such J-dependent shifts have been studied in neon by a parametric approach (Keller, 1973), as already mentioned in Section 3.3.2. He deduced that some of the J-dependent shift was of relativistic origin. J-dependent isotope shifts have been observed in samarium but the question as to whether or not there is a mass shift component of this J-dependent shift will be deferred till Chapter 9.

INTRODUCTION TO THE THEORY OF FIELD SHIFTS IN OPTICAL SPECTRA

4.1. NUCLEAR CHARGE DISTRIBUTIONS

The interaction between atomic electrons (and muons) and nuclei is electromagnetic—there is no strong interaction; the nuclear matter distribution cannot be studied, but the nuclear charge distribution can. The nuclear charge distribution can also be studied by scattering electrons off nuclei elastically. Since the first scattering experiments of Lyman *et al.* (1951) the size of the charge distribution of many nuclei have been found. It has been found that, to a very crude first approximation, nuclei have a uniform charge density ρ, out to a radius R, where there is a sharp edge. The internucleon strong interaction is of such short range that in all but the smallest nuclei it is saturated. This means that nuclear matter has a constant density independent of the number of nucleons in the nucleus A. The radii of all but the lightest nuclei are given by

$$R = 1.2 \, A^{1/3} \, \text{fm} \tag{4.1}$$

A second approximation is that nuclei have a diffuse edge as shown in Fig. 4.1 and the charge density, $\rho(r)$, follows a Fermi distribution. There is really no such thing as *the* nuclear radius; various size parameters or spatial moments of certain distribution functions have to be considered. Fig. 4.1 also shows a more complicated charge distribution function which could be a better approximation for a particular nuclide.

Elastic electron scattering data can give nuclear charge distributions, but the basic problem is that, having calculated a distribution, it is not known whether that distribution is the one that determined the scattering that was observed, or whether it is merely consistent with that observed scattering. This problem is dealt with by Barrett and Jackson (1977) and will be almost entirely ignored in this book. This can be done because isotope shifts do not give detailed information about the nuclear charge distributions, their importance lies in the fact that the limited information which they do give is given with high accuracy for the usually small difference between one isotope and another. A parameter that is very relevant is isotope shift work is the mean square radius of the charge distribution

$$\langle r^2 \rangle = \frac{\int_0^\infty 4\pi r^2 \rho(r) r^2 \, dr}{\int_0^\infty 4\pi r^2 \rho(r) \, dr} = \frac{\int_0^\infty \rho(r) r^4 \, dr}{\int_0^\infty \rho(r) r^2 \, dr} \tag{4.2}$$

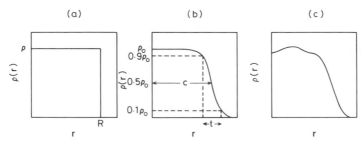

FIGURE 4.1. Nuclear charge distribution functions. (a) A sharp-edged nucleus of constant charge density ρ with a well-defined radius R. (b) The Fermi distribution of the charge density $\rho(r) = \rho_0/\{1 + \exp[(r - c)/a]\}$, where $a = 0.228t$. (c) A possible better approximation for a particular nucleus.

In the very crude first approximation where $\rho(r) = \rho$ from $r = 0$ to $r = R$ and $\rho = 0$ for $r > R$

$$\langle r^2 \rangle = \frac{\int_0^R r^4\, dr}{\int_0^R r^2\, dr} = \frac{3}{5} R^2 \tag{4.3}$$

For the Fermi distribution as defined in Fig. 4.1

$$\langle r^2 \rangle = \tfrac{3}{5} c^2 + \tfrac{7}{5}\pi^2 a^2 \tag{4.4}$$

The mean square radius of the charge distribution of nuclei can be found, for example, from electron scattering data, and multiplication by $5/3$ gives the square of the equivalent uniform radius R_{eq}. Various formulas for R_{eq} have been proposed as an improvement on Eq. (4.1); one which was used by Babushkin (1963) in isotope shift calculations is due to Elton and is

$$R_{eq} = 1.115A^{1/3} + 2.151A^{-1/3} - 1.742A^{-1} \tag{4.5}$$

Such formulas give a reasonable approximation to the truth for $A > 16$, but they do not give the change in nuclear charge radius between isotopes. Indeed, this change could be zero if the extra neutrons of the nucleus of the heavier isotope formed a skin around the protons. To a certain extent this does happen, but in general the protons occupy a larger volume in a heavier isotope so $\langle r^2 \rangle$ increases with A within a set of isotopes.

The field shift involves the electrostatic interaction between the atomic electrons and the nucleus. Outside the nuclear region, the electric field of the nucleus is simply Coulombic due to the nuclear charge Ze, and so is the same for all isotopes; the binding potential is just $-Ze/r$. Within the nuclear

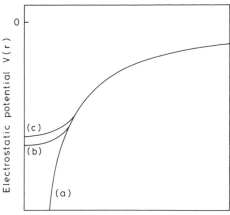

FIGURE 4.2. Electrostatic potentials of nuclei. (a) The Coulomb potential $V = -Ze/4\pi\varepsilon_0 r$ for a point nucleus of charge Ze. (b) The electrostatic potential for a nucleus of the same charge with a finite size of charge distribution. (c) The potential for another nucleus with the same charge but a larger size of charge distribution (an isotope with a larger nucleus, usually a heavier isotope).

region it differs from isotope to isotope, and as can be seen from Fig. 4.2, the larger charge distribution (usually that of the heavier isotope) gives a smaller binding potential. The binding energy or term value of any electron which has a probability of being in the nuclear region is thus smaller for the larger (usually heavier) isotope. Nuclear charge distributions are dealt with in Chapters 9 and 10, but enough has been mentioned here for the purposes of this chapter.

4.2. FIRST-ORDER FIELD SHIFTS

An electronic s–p transition is the paradigm for showing a field shift. The potential difference of Fig. 4.2 has much more effect on the s electron than the p electron, because the s electron density at the nucleus is much larger than the p electron density at the nucleus. The lower level of the transition is thus raised for the larger isotope relative to the smaller; the larger isotope has a smaller transition energy than the smaller isotope. Since the larger isotope is usually the heavier isotope, this means that on the normal convention of heavy minus light, the isotope shift is usually negative.

If the change in electron density at the nucleus during the transition, $\Delta|\psi(0)|^2$, is known, then the potential difference of Fig. 4.2 can be calculated from the isotope shift. This potential difference can be interpreted as a change in the size of the nuclear charge distribution between isotopes. A value for the actual size of the nuclear charge distribution of either isotope can be obtained, but only in a rather indirect way. The transition energy can be calculated, using an *ab initio* theory, for the case of a point nucleus, and the difference between this theoretical value and an experimentally determined transition energy can be used to give the size of the nucleus in question. This technique is of little practical use. On the experimental side it is much more difficult to measure transition energies accurately on an absolute scale than it is to measure differences of energies between isotopes accurately. On the theoretical side it is impossible to calculate transition energies with any confidence except for the simplest hydrogenlike spectra and perhaps the heliumlike ones. In the case of transitions in muonic atoms, both problems can be overcome, hence the importance of work with muonic atoms. However, in the case of optical isotope shifts the information obtained on nuclear sizes is limited to differences between isotopes.

It is easy to see the qualitative connection between a field shift and a change in nuclear size between isotopes. It is much more difficult to make a quantitative connection between the two. This was first done by Racah (1932) and Rosenthal and Breit (1932). The latter's approach has been developed in later work and so it will be briefly described. They considered the relativistic wave function for a one-electron atom with a point nucleus. The effect of a nucleus of finite size was then considered as a perturbation. In making comparison with experimental results, the single electron was taken to be the electron involved in an optical transition in a heavy atom. It was thus a valid approximation to ignore the binding energy of the electron relative to its rest energy. They considered the nuclear charge distribution

$$\rho = \frac{n+1}{4\pi} \frac{Ze}{r_0^3} \left(\frac{r}{r_0} \right)^{n-2} \qquad r < r_0 \qquad (4.6)$$

$$\rho = 0 \qquad\qquad\qquad\qquad r > r_0 \qquad (4.7)$$

when n can vary from -1 for a point charge, through 2 for a uniform charge distribution, to ∞ when all the charge is on the surface. Their first-order perturbation theory result for the field shift of an s electron between nuclei with charge distributions out to r_0 and $r_0 + \delta r_0$ was

$$\delta E_p = 4\pi |\psi(0)|^2 \frac{a_0^3}{Z} R_\infty \frac{(\sigma+1)}{\Gamma^2(2\sigma+1)} \frac{(n+1) y_0^{2\sigma}}{(2\sigma+1)(2\sigma+n+1)} \frac{\delta y_0}{y_0} \quad (4.8)$$

where $|\psi(0)|^2$ is the nonrelativistic probability density of the electron at the point nucleus

$$\sigma^2 = 1 - \alpha^2 Z^2 \qquad (4.9)$$

and

$$y = 2Zr/a_0 \qquad (4.10)$$

the other symbols having their usual meanings (Bohr radius, Rydberg constant, atomic number, gamma function, fine structure constant). They showed that an s electron gave the largest shift and gave an alternative formula for other electrons, for which $\psi(0) = 0$. They pointed out that the nuclear field of a nucleus of finite size distorts the electron wave function quite appreciably and that first-order perturbation theory is inadequate except for nuclei of small Z.

This distortion of the electron wave function was studied more systematically using a nonperturbation method by Broch (1945). The field is less attractive than the Coulomb field within the nucleus and so the electron density within the nucleus is reduced. The change in field between isotopes thus has less effect on the energy of the electron than that given by first-order perturbation theory with unperturbed electron wave functions for a point nucleus. Broch (1945) considered the particular case of all the charge being on the surface, $n = \infty$ in Eq. (4.8), and showed that the first-order perturbation theory result should be multiplied by a factor $2\sigma^2/(\sigma + 1)$.

Bodmer (1953) extended Broch's method to the charge distribution of Eqs. (4.6) and (4.7) and compared the results with the first-order perturbation theory result obtained by Rosenthal and Breit. He found that their result, given as Eq. (4.8), had to be multiplied by a factor P to allow for the distortion of the wave function

$$\delta E = P \delta E_p \qquad (4.11)$$

where

$$P = \frac{2\sigma^2(2\sigma + 1)}{(\sigma + 1)} \frac{2\sigma + n + 1}{n + 1} \frac{1 + (1 + \sigma)K/\alpha Z}{1 + (1 - \sigma)K/\alpha Z} \qquad (4.12)$$

and K was a function of both Z and n, which he calculated for various values.

The charge distributions for $n = 0, 1, 2, 3$, and 4 are shown in Fig. 4.3 and the corresponding values of K were found by Bodmer to be

$$n = 0: \qquad \frac{K}{a} = -\frac{4}{9}(1 + 0.15a^2) \qquad (4.13)$$

$$n = 1: \qquad \frac{K}{a} = -\frac{5}{12}(1 + 0.12a^2 + 0.021a^4) \qquad (4.14)$$

$$n = 2: \qquad \frac{K}{a} = -\frac{2}{5}(1 + 0.106a^2 + 0.0105a^4) \qquad (4.15)$$

$$n = 3: \qquad \frac{K}{a} = -\frac{7}{18}(1 + 0.097a^2 + 0.014a^4) \qquad (4.16)$$

$$n = 4: \qquad \frac{K}{a} = -\frac{8}{21}(1 + 0.092a^2 + 0.0135a^4) \qquad (4.17)$$

$$n = \infty: \qquad \frac{K}{a} = -\frac{1}{3}(1 + 0.067a^2 + 0.00635a^4) \qquad (4.18)$$

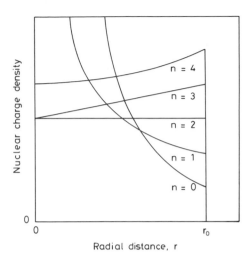

FIGURE 4.3. The nuclear charge distributions according to Eq. (4.6) for various values of n.

where

$$a = \alpha Z \tag{4.19}$$

For charge distributions given by Eqs. (4.6) and (4.7)

$$\langle r^2 \rangle = \frac{n+1}{n+3} r_0^2 \tag{4.20}$$

and so it follows from Eqs. (4.8)–(4.20) that

$$\delta E = \pi |\psi(0)|^2 \frac{a_0^3}{Z} \zeta N \langle r^2 \rangle^{\sigma-1} \delta\langle r^2 \rangle \tag{4.21}$$

where ζ is a function of Z

$$\zeta = R_\infty \left(\frac{2Z}{a_0} \right)^{2\sigma} \frac{1}{\Gamma^2(2\sigma)} \tag{4.22}$$

and N is a function of Z and n

$$N = \left(\frac{n+3}{n+1} \right)^\sigma \frac{1+(1+\sigma)K/\alpha Z}{1+(1-\sigma)K/\alpha Z} \tag{4.23}$$

In fact the variation of N with n for a given Z is extremely weak as can be seen from Table 4.1. If this weak variation with n is neglected then, as Bodmer (1959) stressed, the relevant property of the nuclear charge distribution so far as isotope shifts is concerned is the mean square radius $\langle r^2 \rangle$, defined in Eq. (4.2). This appears in Eq. (4.21) as $\langle r^2 \rangle^{\sigma-1} \delta\langle r^2 \rangle$. Now as σ is

TABLE 4.1
Variation of Field Shift with Nuclear Charge Distribution

n	Values of N given by Eq. (4.23) (atomic number, Z)				
Eq. (4.6)	41	55	68	82	96
0	0.349	0.364	0.385	0.417	0.462
1	0.350	0.363	0.383	0.411	0.451
2	0.350	0.364	0.385	0.413	0.454
3	0.350	0.364	0.384	0.413	0.453
4	0.350	0.364	0.384	0.413	0.453
∞	0.350	0.365	0.385	0.413	0.454

of the order of 0.8 or 0.9 for heavy elements, the term $\langle r^2 \rangle^{\sigma-1}$ varies only slowly with $\langle r^2 \rangle$, and for the lighter elements it tends to unity. An approximate value of $\langle r^2 \rangle^{\sigma-1}$ is all that is needed for substitution into Eq. (4.21); $\langle r^2 \rangle = 3R^2/5$, where R is given by Eq. (4.5), is quite adequate, and indeed for the lighter elements only Eq. (4.1) need be used. Eq. (4.21) can thus be written as

$$\delta E = \pi |\psi(0)|^2 \frac{a_0^3}{Z} f(Z) \delta \langle r^2 \rangle \qquad (4.24)$$

where $f(Z)$ is a known function of Z (and A and n, to be precise). It is written $f(Z)$ to follow the usage of a much quoted paper by Heilig and Steudel (1974). It is also customary, following Brix and Kopfermann (1949), to refer to

$$f(Z) \delta \langle r^2 \rangle = C \qquad (4.25)$$

as the isotope shift constant. In terms of this Eq. (4.21) becomes

$$\delta E = \pi |\psi(0)|^2 \frac{a_0^3}{Z} C \qquad (4.26)$$

with

$$C = \zeta N \langle r^2 \rangle^{\sigma-1} \delta \langle r^2 \rangle \qquad (4.27)$$

Fradkin (1962) also considered the charge distributions of Eqs. (4.6) and (4.7) and, like Bodmer, concluded that the field shift had only a weak dependence on the details of the nuclear charge distribution.

Babushkin (1963) determined the required wave functions by a different method which obviated the need to assume that the electron binding energy was negligible. He tabulated calculated values of the isotope shift constant for two particular uniform nuclear charge distributions. To make these figures

TABLE 4.2
Field Shift Functions from the Calculations of Isotope Shift Constants by Babushkin (1963)[a]

Element	Z	ζN (GHz fm$^{-2\sigma}$)	$f(Z)$ (GHz fm^{-2})	Element	Z	ζN (GHz fm$^{-2\sigma}$)	$f(Z)$ (GHz fm^{-2})
Rb	37	3.79	3.41	Eu	63	29.70	20.79
Sr	38	4.11	3.67	Gd	64	32.43	22.41
Ru	44	6.69	5.73	Yb	70	52.90	33.50
Pd	46	7.82	6.55	Hf	72	62.62	38.36
Ag	47	8.46	7.08	W	74	74.06	43.95
Cd	48	9.17	7.58	Re	75	80.8	47.20
Sn	50	10.70	8.68	Os	76	87.8	50.34
Xe	54	14.63	11.37	Ir	77	95.8	53.94
Ba	56	17.12	13.05	Pt	78	104.4	57.7
Ce	58	20.10	14.98	Hg	80	124.2	66.1
Nd	60	23.58	17.13	Tl	81	135.8	70.9
Sm	62	27.59	19.58	Pb	82	148.4	76.0

[a]These figures were obtained from Babushkin's with the aid of Eq. (4.25) for $f(Z)$ and Eq. (4.27) for ζN.

more generally applicable they have been converted, in Table 4.2, into values of ζN and $f(Z)$. In the latter case, Elton's values of R of Eq. (4.5) were used to calculate $\langle r^2 \rangle^{\sigma-1}$. Figures for other values of Z can be obtained by interpolation. Apart from any other errors, there is a rounding error of about 2 in the last figure for both ζN and $f(Z)$. Since 1963, interest in field shifts has extended to lighter elements than rubidium and figures for some of these are given in Table 4.3. Eqs. (4.22) and (4.23) were used with slight alterations to the heavier elements to make the figures merge smoothly with those of Table 4.2. The field shift is not exactly proportional to $\langle r^2 \rangle^{\sigma-1}\delta\langle r^2 \rangle$, but in view of the uncertainties involved in finding the change in the electron density at the nucleus during an optical transition, $\Delta|\psi(0)|^2$, it is a good enough approximation for optical isotope shift work. A more exact theory (Seltzer, 1969) which should be used for the very heaviest elements is mentioned in Section 5.1.

4.2.1. THE ELECTRONIC FACTOR

So far, only the nuclear factor of Eq. (4.21) has been considered. The electronic factor of Eq. (4.21) is $\pi|\psi(0)|^2 a_0^3/Z$, the theory assuming that the field shift was caused by one electron with a density at the point nucleus of $|\psi(0)|^2$. Isotope shifts are actually measured in transitions between levels which both have values of $|\psi(0)|^2$ and it is $\Delta|\psi(0)|^2$ that matters, the change in $|\psi(0)|^2$ during the transition. The determination of this will be dealt with in Section 4.2.2. An approximation to the one-electron case is an $ns - n'p_{\frac{3}{2}}$ transition in an alkali-like spectrum. Such transitions are often studied in

TABLE 4.3
Field Shift Functions for Lighter Elements

Element	Z	ζN (GHz fm$^{-2\sigma}$)	$f(Z)$ (GHz fm^{-2})	Element	Z	ζN (GHz fm$^{-2\sigma}$)	$f(Z)$ (GHz fm^{-2})
H	1	0.0016	0.0016	V	23	1.05	1.01
He	2	0.0063	0.0063	Cr	24	1.16	1.12
Li	3	0.0142	0.0142	Mn	25	1.29	1.23
Be	4	0.0253	0.0253	Fe	26	1.42	1.36
B	5	0.0397	0.0396	Co	27	1.57	1.49
C	6	0.0574	0.0573	Ni	28	1.72	1.63
N	7	0.0787	0.0784	Cu	29	1.89	1.78
O	8	0.103	0.103	Zn	30	2.07	1.94
F	9	0.132	0.132	Ga	31	2.26	2.11
Ne	10	0.164	0.164	Ge	32	2.47	2.29
Na	11	0.201	0.200	As	33	2.70	2.49
Mg	12	0.242	0.240	Se	34	2.94	2.70
Al	13	0.287	0.284	Br	35	3.20	2.92
Si	14	0.337	0.333	Kr	36	3.48	3.16
P	15	0.392	0.386	Rb	37	3.79	3.41
S	16	0.452	0.445	Sr	38	4.11	3.67
Cl	17	0.518	0.508	Y	39	4.47	3.97
Ar	18	0.590	0.577	Zr	40	4.85	4.28
K	19	0.668	0.652	Nb	41	5.26	4.61
Ca	20	0.752	0.732	Mo	42	5.70	4.97
Sc	21	0.844	0.818	Tc	43	6.18	5.34
Ti	22	0.942	0.911	Ru	44	6.69	5.73

isotope shift work because the ns electron has a large density at the nucleus, and the $n'p_{\frac{3}{2}}$ electron a very small one (even smaller than that of an $n'p_{\frac{1}{2}}$ electron according to relativistic theory).

The density at the point nucleus of the ns electron alone, $|\psi(0)|^2(ns)$, can be found either by an *ab initio* calculation or by an empirical method. There are in fact two closely related empirical methods: one involves the Goudsmit–Fermi–Segrè formula, the other the magnetic hyperfine splitting for a nucleus of known spin and magnetic dipole moment. Considering the latter first, all the inner electrons are paired off in an alkali-like atom and make no contribution to the magnetic hyperfine splitting. The splitting is directly proportional to the density at the point nucleus of the ns electron. The connection is (see for example, Kopfermann and Schneider, 1958)

$$\pi|\psi(0)|^2(ns)a_0^3 = \frac{A_{ns}\sigma(4\sigma^2 - 1)}{8hcR_\infty\alpha^2(1 - \delta)(1 - \varepsilon)g_I'} \tag{4.28}$$

where A_{ns} is the magnetic hyperfine splitting factor of the ns electron and δ is a correction for the finite size of the nucleus (Crawford and Schawlow, 1949).

Only a very approximate value of the nuclear size is required, and so δ can be treated as a function of Z (see Table 4.4). ε is a correction for the finite size over which the nuclear magnetic moment is distributed (Bohr and Weisskopf, 1950); in general, ε is also a function of Z going from zero to 0.03 as Z increases. Its value is, however, specific to different isotopes and it can vary from the general range of values given above. Care must be taken when this happens, but the effect shows itself as what is known as the hyperfine structure anomaly and so should be apparent when using hyperfine structure data for the determination of $|\psi(0)|^2(ns)$. g_I' is the reduced nuclear g factor

$$g_I' = \frac{\mu_I/\mu_B}{I} \tag{4.29}$$

where μ_I is the magnetic moment of the nucleus, μ_B is the Bohr magneton, and I is the nuclear spin quantum number.

Bauche (1981) has pointed out that there is some evidence that the magnetic hyperfine splitting and the isotope (field) shift do not depend on the nonrelativistic $|\psi(0)|^2(ns)$ in the same way. In particular, the relativistic corrections which have to be applied are not the same in the two effects. The field shift depends on the electron wave function only in the nucleus, whereas the magnetic hyperfine splitting contains a radial $[r^{-2}]$ integral in which larger values of r are important. He suggests that if results to better than 5% are required, it will be necessary to do more work on the comparison of the relativistic with the nonrelativistic theories of the two effects.

The other, closely related, empirical method of determining $|\psi(0)|^2(ns)$ is by the use of the Goudsmit–Fermi–Segrè formula. This links the s electron density at the nucleus with the effective quantum number of the ns level. Goudsmit (1933) and Fermi and Segrè (1933) found that the formula for a hydrogenlike atom with principal quantum number n

$$\pi|\psi(0)|^2(ns)a_0^3 = \frac{Z^3}{n^3} \tag{4.30}$$

could be modified, for an s level in an alkali-like atom, to

$$\pi|\psi(0)|^2(ns)a_0^3 = \frac{ZZ_a^2}{n*^3}\frac{dn*}{dn} \tag{4.31}$$

TABLE 4.4
Values of δ in Eq. (4.28)

Z	20	30	40	50	60	70	75	80	85	90
δ	0.002	0.004	0.011	0.025	0.044	0.071	0.10	0.13	0.17	0.21

where Z_a is the spectrum number (1 for neutral atoms, 2 for singly-ionized atoms, etc.), and n^* is the effective quantum number. The formula has turned out to be surprisingly, even suspiciously, accurate and has been reconsidered by Foldy (1958) and Iwinski *et al.* (1980). The latter have generalized it to arbitrary angular momentum and to inner shell electrons. They show that Z_a, the suspiciously arbitrary factor in the formula, is not intrinsically present, but occurs only because of the use of the effective quantum number.

The *ab initio* calculation of $|\psi(0)|^2(ns)$ for the prediction of isotope shifts in heavy elements was first carried out by Wilson (1968). Such Hartree–Fock calculations do not give very precise values of the ns electron density at the nucleus, and the empirical methods already described for finding the density are to be preferred. These calculations also made explicit what had already been suspected; that the single electron approach is not a good approximation to the subject of field shifts in alkali-like atoms. In an ns–n'p transition, the screening of all the other s electrons in the atom is altered and so their densities at the nucleus are altered. It may be that $|\psi(0)|^2(ns)$ is a poor approximation to $\Delta|\psi(0)|^2(ns$–n'p$)$, the change in the total electron density at the nucleus during the transition, which is the relevant quantity. This point will be taken up in Section 4.2.2.

A better approach to the theory of field shifts is to develop a relativistic theory for the atom as a whole, and to determine the change in the electron–nuclear interaction during the transition. This involves finding the difference between two large and nearly equal numbers, a notoriously difficult operation which has been compared with finding the weight of a captain by weighing his ship with him on board and on shore. It has, however, been attempted by Grant (1980) who had enough computing power available to reduce the truncation errors to a reasonable size. Although the technology of computing relativistic self-consistent fields is adequate for the estimation of isotope shifts, the method needs further development. As it stands, it predicts that non-s electrons make large contributions to the field shift. Experimental results on the other hand can be convincingly explained on the assumption that the dominant contribution of non-s electrons is by their screening of the s electrons. Unfortunately, there has so far been no clear cut case which can distinguish between the two approaches.

At present the best approach seems to be to start with the single electron relativistic theory and then make allowance for the screening effect of the other electrons. These screening effects will be considered in the next section, but this section will be concluded by considering the particular example of the alkali-like spectrum Ba II.

Isotope shifts have been measured in the transition $6s\,^2S_{1/2}$–$6p\,^2P_{3/2}$ (Fischer *et al.*, 1974a) and so the electron density of interest is $|\psi(0)|^2(6s)$. Although the single-electron theory is relativistic, it is usually expressed in terms of the nonrelativistic $|\psi(0)|^2$, which would be obtained with a point nucleus. The magnetic hyperfine splitting has been measured and from Eq. (4.28) it is found that $\pi|\psi(0)|^2(6s)a_0^3$ is 18.9. The use of the Goudsmit–Fermi–Segrè formula leads to a value of 19.2. An *ab initio* value has been calculated by Wilson (1978a) using Hartree–Fock wave functions, the result being 12.8. There is a further, if very roundabout, way of obtaining a value in the case of barium, since the sizes of the nuclear charge distributions have been determined from the spectra of muonic barium atoms. It is thus possible to work backwards from a known $\delta\langle r^2\rangle$ via a knowledge of the isotope shifts and the screening ratios to obtain the value 15.2 for $\pi|\psi(0)|^2(6s)a_0^3$.

This result is not completely independent of the *ab initio* result of 12.8 already quoted because the screening factors have to be calculated using the same *ab initio* calculations. However, as will be explained in the next section, the *ab initio* calculations give much more accurate values of screening ratios than they do of values of $\pi|\psi(0)|^2a_0^3$ so the result of 15.2 is virtually independent of the *ab initio* result of 12.8. Such Hartree–Fock calculations are not expected to give accurate absolute values, so the discrepancy between 12.8 and the other results is not a cause for concern. The remaining discrepancy, between 15.2 and 18.9 and 19.2, perhaps is. It certainly shows that we have hardly reached the 5% level mentioned by Bauche (1981).

The steps in obtaining the value of 15.2 are, very briefly, as follows: for ^{138,136}Ba, $\delta\langle r^2\rangle = 0.044$ fm^2 (Section 9.56); the mass shift in $\lambda 553.5$ nm is 9 MHz (Section 9.56); it follows that the mass shift in $\lambda 493.4$ nm, the $6s\,^2S_{1/2}$–$6p\,^2P_{3/2}$ transition of Ba II, is -12 MHz (Fig. 6.3); the field shift is -173 MHz; for the 6s–6p transition, $\pi\Delta|\psi(0)|^2a_0^3 = 16.9$ (Table 4.2); the screening ratio $\Delta|\psi(0)|^2(6s\text{–}6p)/|\psi(0)|^2(6s) = 1.11$ (Wilson, 1978a), hence $|\psi(0)|^2(6s) = 16.9/1.11 = 15.2$. See also Fig. 9.7 of Section 9.56.

4.2.2. SCREENING EFFECTS

The size of the field shift in a transition is proportional to the change in the total electron density at the nucleus. To a first approximation, this change is the change in the density at the nucleus of the transition electron. In addition, however, the density of the other electrons may change because of the change in the transition electron. This is because the electrons are not independent but screen each other. This effect was pointed out by Breit (1932) and considered quantitatively by Crawford and Schawlow (1949). Experimental evidence for such screening effects was assembled by Brix and Kopfermann (1958) for the first and second spectra of gadolinium. They found, for example, that the shift in d^2p–d^2s of Gd I was smaller than that in dp–ds of Gd II; for this and other transitions the extra 5d electron of Gd I reduced the shift of the 6s electron by about 25%. They also found that the shift in s^2p–dsp was not as large as that in sp–dp; the second 5s electron screened the first one, again by about 25%. Very similar values had been obtained for mercury and it was obvious that the outer electrons were screening each other, and that the amount of screening changed during a transition. It was also realized that a further uncertainty existed, the inner closed shells, $1s^2$, $2s^2$, etc., could also be affected by changes in the configuration of the outer electrons. These screening effects can be accounted for by inserting a screening factor β, into Eq. (4.26) and other similar equations

$$\delta E = \pi\beta|\psi(0)|^2 \frac{a_0^3}{Z} C \qquad (4.32)$$

This allows for the fact that the quantity which is really involved is the change in the total electron density at the nucleus during the transition, $\Delta|\psi(0)|^2$, rather than the density at the nucleus of one ns electron, $|\psi(0)|^2(ns)$. The latter is useful in developing the theory and in determining the electronic factor as discussed in Section 4.2.1, but the screening effects must also be taken into account

$$\Delta|\psi(0)|^2(\text{all electrons}) = \beta|\psi(0)|^2(ns) \qquad (4.33)$$

If the screening factor is not known, then values of βC rather than C, or $\beta\delta\langle r^2\rangle$ rather than $\delta\langle r^2\rangle$, are all that can be determined.

It was not until 1968 that any realistic evaluations of β were made. Wilson (1968) made Hartree–Fock calculations for platinum, mercury, and thallium; some of his results are given in Table 4.5 for mercury. From these results, it is possible to determine screening ratios which can be compared with experiment. As explained in Section 6.1 it is possible to determine the ratio of the field shifts measured in two transitions of the same element; the effect of the mass shifts can be eliminated exactly. Such experimentally determined ratios can be compared with the corresponding calculated ratios of $\Delta|\psi(0)|^2$, the screening ratios.

For example, it should be found that

$$\frac{FS(5d^{10}6s^2-5d^{10}6s6p)}{FS(5d^96s^2-5d^96s6p)} = \frac{\Delta|\psi(0)|^2(5d^{10}6s^2-5d^{10}6s6p)}{\Delta|\psi(0)|^2(5d^96s^2-5d^96s6p)} \qquad (4.34)$$

which according to Table 4.5 is equal to $143/234 = 0.61$. A comparison of such experimental field shift ratios with calculated screening ratios has shown that the calculated screening ratios are fairly reliable. This gives one confidence in using the calculated values of electron density at the nucleus to determine screening factors even though these cannot be checked directly. For example, if the isotope shift is measured in $5d^{10}6s^2-5d^{10}6s6p$ then it follows from Table 4.5 that the screening factor

$$\beta = \frac{\Delta|\psi(0)|^2(5d^{10}6s^2-5d^{10}6s6p)}{|\psi(0)|^2(6s \text{ in } 5d^{10}6s)} = \frac{143}{192} = 0.74 \qquad (4.35)$$

TABLE 4.5
Hartree–Fock Values of $4\pi|\psi(0)|^2 a_0^3$ for Mercury[a]

Contributing electrons	Configuration of Hg I		
	$5d^{10}6s^2$	$5d^{10}6s6p$	$5d^9 6s^2 6p$
$1s^2$	4,039,222	4,039,226	4,039,222
$2s^2$	448,405	448,405	448,408
$3s^2$	101,053	101,052	101,059
$4s^2$	25,101	25,099	25,105
$5s^2$	4,597	4,587	4,652
6s or $6s^2$	300	167	389
Total	4,618,678	4,618,536	4,618,835

Contributing electrons	Configuration of Hg II		
	$5d^{10}6s$	$5d^9 6s^2$	$5d^9 6s6p$
$1s^2$	4,039,230	4,039,228	4,039,224
$2s^2$	448,406	448,408	448,406
$3s^2$	101,053	101,059	101,056
$4s^2$	25,099	25,106	25,102
$5s^2$	4,586	4,653	4,640
6s or $6s^2$	192	442	233
Total	4,618,566	4,618,896	4,618,662

[a]The figures are from Wilson (1968).

so, as the hyperfine structure of 6s in $5d^{10}6s$ is known, it is possible to determine an empirical value of $|\psi(0)|^2$(6s in $5d^{10}6s$) and hence an empirical value of $\Delta|\psi(0)|^2$. Such indirectly determined values of $\Delta|\psi(0)|^2$ are found to be more accurate than the directly calculated values. The Hartree–Fock calculations give reasonably good estimates of ratios of electron densities but not nearly such good absolute values.

Hartree–Fock calculations of screening ratios are not always in good agreement with experimentally determined field shift ratios, as was pointed out by Wilson (1972) in the case of the rare earths, samarium and europium. An obvious explanation of such discrepancies is that in complicated spectra there is much configuration mixing, whereas the calculations were for single, pure, undistorted configurations. Wilson pointed out that relativistic effects might also be a cause, since (see, for example: Grant, 1970) the relativistic effects vary with the angular momentum of the electron. Coulthard (1973) applied his relativistic Hartree–Fock atomic structure computer program to the calculation of screening ratios in europium. He obtained relativistic enhancements of $|\psi(0)|^2$ of from three to four which led to considerable changes in the screening ratio for a few transitions. To take an extreme example

$$\frac{\Delta|\psi(0)|^2(f^6ds^2-f^7s^2)}{\Delta|\psi(0)|^2(f^7s^2-f^7sp)} = \begin{cases} 2.94 \text{ (Wilson, nonrelativistic)} \\ 2.31 \text{ (Coulthard, relativistic)} \end{cases} \qquad (4.36)$$

showing that relativistic effects were not negligible. The comparison of such ratios with experimental values is difficult because of the lack of pure configurations in the rare earth spectra. In the case of neodymium (King *et al.*, 1973) the calculated screening ratios were used to estimate the amount of configuration mixing in some impure levels. The self-consistency obtained was slightly better for relativistic than for nonrelativistic calculations. The relativistic Hartree–Fock calculations are very complicated, and to include configuration mixing and variations between terms of a configuration, as one should in a relativistic treatment, would be in Coulthard's words, "a considerable undertaking."

A simpler method of obtaining some of the relativistic corrections to atomic radial wave functions was suggested by Cowan and Griffin (1976). Their method, which they called the HFR method, incorporated the major relativistic effects within the format of the nonrelativistic approach. The mass–velocity and Darwin terms of the Pauli equation for one-electron atoms are added to the usual nonrelativistic one-electron differential equation for the radial function. The computation is much simpler than in the fully relativistic Dirac–Hartree–Fock method, but $p_{1/2}$ and $p_{3/2}$ electrons are not distinguished so the finite electron density at the nucleus of the former is neglected. This HFR method was applied to the calculation of screening ratios in barium (Wilson, 1978a) where a comparison was made with nonrelativistic ratios. For the ratio (5d–6p)/(6s–6p) in Ba II, the pseudo-relativistic value is -0.23 (see Table 9.11), and the nonrelativistic Hartree–Fock value is -0.32. The experimentally determined ratio of the field shifts (Höhle *et al.*, 1978), -0.22 (see Section 9.56) shows that the pseudo-relativistic calculation is much to be preferred.

The screening factor in Ba II

$$\beta = \frac{\Delta|\psi(0)|^2(5p^66s-5p^66p)}{|\psi(0)|^2(6s \text{ in } 5p^66s)} = \begin{cases} \dfrac{128}{115} = 1.11 \text{ (HFR)} \\ \dfrac{57.0}{51.4} = 1.11 \text{ (HF)} \end{cases} \qquad (4.37)$$

is not affected by relativistic corrections and so can be used with confidence to derive values of $\delta\langle r^2 \rangle$ with the aid of empirically determined values of $|\psi(0)|^2(6s)$. The results of such calculations are given in Section 9.56.

4.2.3. FIELD SHIFTS OF NONSPHERICAL NUCLEI

It was first pointed out by Brix and Kopfermann (1949), when discussing isotope shifts in samarium, that a change in the shape of a nucleus can give rise to a field shift even if the actual volume of the nuclear charge distribution does not change. Thus the change in $\langle r^2 \rangle$ between isotopes, $\delta \langle r^2 \rangle$, can arise from changes in nuclear shape as well as in nuclear volume. The analyses of experiments to find the charge distributions of deformed nuclei are very dependent on the model used for the nuclear charge distribution. The results obtained by different techniques (muonic hyperfine spectra, inelastic electron scattering, and Coulomb excitation) are in good agreement and so it is possible to use the results to apportion the $\delta \langle r^2 \rangle$ obtained from isotope shifts into a volume component and a shape component.

Consider the simplest model of the nuclear charge distribution—that the charge distribution is uniform within the sharp nuclear surface. For a deformed nucleus this surface is usually described in terms of spherical harmonics, the most important one being

$$Y_2^0 = \left(\frac{5}{16\pi} \right)^{1/2} (3\cos^2\theta - 1) \tag{4.38}$$

The coefficient of this, β_2^0, gives the amount of this type of deformation, a quadrupole deformation with axial symmetry, θ being the angle from the axis. A positive value of β_2^0 gives a prolate (rugby ball) shape. So for this type of deformation

$$r = r_0 \left(1 + \beta_2 Y_2^0 \right) \tag{4.39}$$

where the superscript on β_2^0 has been omitted, and will be from now on. For this surface shape, and for terms up to the second power of β_2

$$\langle r^2 \rangle \simeq \frac{3}{5} r_0^2 \left(1 + \frac{7}{4\pi} \beta_2^2 \right) \tag{4.40}$$

and the volume

$$V \simeq \frac{4}{3} \pi r_0^3 \left(1 + \frac{3}{4\pi} \beta_2^2 \right) \tag{4.41}$$

Obviously $\langle r^2 \rangle$ will change with β_2^2 even if V is kept fixed, giving rise to a shape component of $\delta \langle r^2 \rangle$.

Consider the case where this effect was first noted—samarium. We now know the values of $\langle r^2 \rangle$ and β_2^2 for ^{154}Sm, ^{152}Sm, and ^{150}Sm. Using the above simple model (uniform charge distribution out to an axially symmetric

TABLE 4.6
Deformations and Sizes of Some Samarium Nuclei

	^{150}Sm	^{a}Sm	^{152}Sm	^{b}Sm	^{154}Sm
β_2^2	0.032	0.082	0.082	0.100	0.100
$\langle r^2 \rangle / \text{fm}^2$	25.30	25.79	25.72	25.89	25.94
r_0^2 / fm^2	41.43	41.10	40.99	40.87	40.95
V / fm^3	1126	1126	1121	1121	1124

^{a}The hypothetical nuclide with the volume of ^{150}Sm and the deformation of ^{152}Sm.
^{b}The hypothetical nuclide with the volume of ^{152}Sm and the deformation of ^{154}Sm.

quadrupole deformed surface), $\langle r^2 \rangle$ can be calculated for two hypothetical nuclides with the volume of a lighter isotope and the deformation of a heavier isotope. The results are shown in Table 4.6, and it can be seen that the changes in β_2^2 have more effect on $\langle r^2 \rangle$ than the changes in volume; in this example the shape components are larger than the volume components. Indeed, between ^{152}Sm and ^{150}Sm the volume component is negative. Even between ^{154}Sm and ^{152}Sm the shape component is much more than the volume component, $\delta \langle r^2 \rangle_{\text{shape}} = 0.17 \text{ fm}^2$ and $\delta \langle r^2 \rangle_{\text{volume}} = 0.05 \text{ fm}^2$.

The shape component can be found from Eqs. (4.40) and (4.41) by allowing β_2^2 to vary while keeping V constant;

$$\frac{\partial \langle r^2 \rangle}{\partial (\beta_2^2)} = \frac{5}{4\pi} \langle r^2 \rangle \left(1 - \frac{5}{4\pi} \beta_2^2 \right) \tag{4.42}$$

for terms up to β_2^2, where the mean square radius of the nuclear charge distribution and the deformation are the averages between the two isotopes involved.

Insofar as the field shift is proportional to $\delta \langle r^2 \rangle$ and does not otherwise depend on the nuclear size, the field shift can be split into two components, the volume shift and the shape shift, just as $\delta \langle r^2 \rangle$ has been. This is not strictly true as can be seen from Eq. (4.21), and real nuclei do not have uniform charge distributions. If allowance is made for these complications then the equations for handling nuclear deformation become very cumbersome (see, for example: Stacey, 1966), but these more general results are very rarely used in practice.

Values of β_2^2 for the two isotopes and hence $\delta(\beta_2^2)$ are usually obtained from Coulomb excitation experiments, inelastic electron scattering experiments, or the study of muonic hyperfine structure. The results of such experiments are often expressed in terms of the intrinsic quadrupole moment

$$Q = \frac{3}{(5\pi)^{1/2}} Z r_0^2 \beta_2 (1 + 0.36\beta_2) \tag{4.43}$$

where it should be borne in mind that a nucleus of spin zero which behaves as a spherical nucleus may nevertheless possess a finite intrinsic quadrupole moment.

In Coulomb excitation a charged particle such as a proton or α-particle is scattered by a nucleus, but the particle is of sufficiently low energy that nuclear, as opposed to electromagnetic, forces may be disregarded. The interaction involves the charge and current distributions in the nucleus, and the transition probability of an electric quadrupole transition of the nucleus from its ground state to an excited state involves the intrinsic quadrupole moment of the ground state. The part of the transition probability which contains nuclear information is called the reduced transition probability. For even–even nuclei with no angular momentum in the ground state, the reduced transition probability for an electric quadrupole transition from the ground state is related to the deformation parameter β_2, of the ground state by

$$\underset{0 \to 2}{B(E2)} = \left(\frac{3}{4\pi} Zer_0^2 \right)^2 \beta_2^2 \qquad (4.44)$$

It was such Coulomb excitation experiments which gave the values of β_2^2 used in Table 4.6.

Some nuclei have equally spaced excited states which are characteristic of vibrational states. These states of nuclei are formed by flexings of the nuclear surface about a spherical shape and are observed, for example, in $^{114}_{48}$Cd and $^{152}_{64}$Gd and their neighboring nuclides. For such vibrational nuclei, the mean value of β_2, $\langle \beta_2 \rangle$ is zero, but the mean value of β_2^2, $\langle \beta_2^2 \rangle$ is not. The latter is the relevant quantity for both Coulomb excitation experiments, Eq. (4.44), and isotope shift measurements, Eq. (4.40), when vibrational nuclei are under consideration.

4.3. SECOND-ORDER FIELD SHIFTS

So far, in this chapter, the field shift has been discussed in terms of electron configurations which give the first-order field shift. With the advent of very high resolution spectroscopy, it has become possible to show that the size of the field shift of a configuration varies with the term or even the level of the configuration. These variations arise from second-order contributions and from relativistic effects. The matter is discussed in the review article of Bauche and Champeau (1976) and these ideas have been applied in the parametric method of analyzing field shift measurements.

Aufmuth (1982) has considered the theory of second-order field shifts in the case of an s electron coupled to a core of Nl (non-s) electrons, i.e., the configuration l^Ns. The second-order contribution involves all other configurations where an s electron (either the valence or a core electron) is excited into another s configuration. Since these configurations usually lie high in the level scheme, the effects are also called far-configuration-mixing effects. Aufmuth showed that *ab initio* calculations of these effects could be made with the Hartree–Fock method. He found that the effect was largest for a configuration in the middle of a shell. A good example is molybdenum (Aufmuth *et al.*, 1978) which has a partly filled 4d shell. The configuration of interest is $4d^5 5s$ of Mo I; the isotope shift has been measured between the two levels $^7S_3$ and $^5S_2$ of this configuration, and common upper levels of the configuration $4d^5 5p$. The two field shifts, instead of being very similar as would be expected on first-order field shift theory, were found to differ by a factor of about two. After allowance had been made for mass shifts and for configuration mixing, it was found that the field shifts for ^{94,92}Mo were -660 MHz in the transition $4d^5 5s \, ^7S_3 - 4d^5 5p \, ^7P$, and -380 MHz in $4d^5 5s \, ^5S_2 - 4d^5 5p \, ^7P$. The difference gives the field shift in $4d^5 5s \, ^7S_3 - ^5S_2$ to be -280 MHz, or -310 MHz according to Aufmuth (1982), with an experimental uncertainty of only about 90 MHz. This difference must arise from the second-order effects and Aufmuth (1982) has calculated a value for these from theory of -320 MHz. The agreement with the experimental value is thus excellent. Aufmuth (1982) gives many other comparisons of theory with experiment, but in most cases the latter is liable to large uncertainties. This is because the need for a common upper level involves the investigation of a weak intercombination line; in the molybdenum example, this is $4d^5 5s \, ^5S_2 - 4d^5 5p \, ^7P_2$. At the time of writing, deliberate efforts are being made to measure isotope shifts in such lines, and results should soon be published for the spectra of zirconium, ruthenium, tungsten, and osmium.

4.4. FIELD SHIFTS OF NON-S ELECTRONS

As already explained in Section 4.2.1, the field shift in a transition is proportional to $\Delta|\psi(0)|^2$, the change in the electron density at the nucleus. This quantity is primarily determined by the s electrons, but non-s electrons do make smaller contributions, both directly and indirectly.

The direct contribution is that, on a relativistic single electron theory as used in Section 4.2, the electron density at the nucleus is finite for $p_{1/2}$ electrons as well as for s electrons, although much smaller for the former. Fradkin (1962) gives the ratio of the charge densities at the nucleus, and hence of the field shifts, as

$$\frac{|\psi(0)|^2(n\mathrm{p}_{1/2})}{|\psi(0)|^2(n\mathrm{s})} = \left(\frac{\alpha Z}{1+\sigma} \right)^2 \tag{4.45}$$

TABLE 4.7
Charge Density at the Nucleus of $p_{1/2}$ Electrons

Element	Z	αZ	σ	$\dfrac{\|\psi(0)\|^2(np_{1/2})}{\|\psi(0)\|^2(ns)}$[a]
Rb	37	0.270	0.963	0.019
Ag	47	0.343	0.939	0.031
Cs	55	0.401	0.916	0.044
Au	79	0.577	0.817	0.101
Fr	87	0.635	0.773	0.128

[a] See Eq. (4.45), which is taken from Fradkin (1962).

where n is the principal quantum number and the other symbols have their usual meanings [see Eqs. (4.9) and (4.10)]. Some values of this ratio are given in Table 4.7 for elements in which a single electron approach is not unreasonable. It can be seen that the contribution of a $p_{1/2}$ electron in a heavy atom can be over 10% of that of an s electron with the same principal quantum number. According to the relativistic theory of Grant (1980), non-s electrons can make even larger contributions but, as mentioned in Section 4.2.1, there is as yet no experimental confirmation of this.

The indirect contribution is that non-s electrons screen s electrons, and so alter the charge density of the latter at the nucleus. This is discussed in Section 4.2.2.

ISOTOPE SHIFTS IN X-RAY SPECTRA

5.1. ELECTRONIC ATOMS

In an electronic, as opposed to a muonic, atom an x-ray transition can exhibit an isotope shift for the same reasons that an optical transition can. Compared with optical transitions, the interpretation of shifts in x-ray transitions is simpler but unfortunately their measurement with precision is not. In practice, the only x-ray transitions in which shifts can be usefully measured are the K lines, and the one invariably studied is the $K\alpha_1$ line ($1s_{1/2}-2p_{3/2}$). The theoretical interpretation is simpler than that of optical transitions because only closed shell or single hole configurations are involved. The variation from element to element is smooth and gradual; each element and each stage of ionization is not completely different as in optical spectra.

The normal mass shift is about 2 meV for $\delta N = 2$ for all heavy elements. It is almost independent of Z because as a fraction of the transition energy it is, according to Eq. (3.4), proportional to M^{-2}. As the $K\alpha$ line involves the transition of an innermost electron, the energy of the transition is proportional to $(Z-s)^2$ where the screening effect s, is very small. Thus, insofar as M is approximately proportional to Z, the normal mass shift in $K\alpha$ transitions is independent of Z. The specific mass shift has been evaluated using Vinti's method with Herman and Skillman wavefunctions by Chesler and Boehm (1968). They found it depended essentially on only the core electrons and so varied smoothly with Z. For $Z > 40$ it is $-1/3$ times the normal mass shift. The total mass shift is thus 1.3 meV for $\delta N = 2$ in $K\alpha$ lines. Any variation between elements can be ignored as the experimental error in measured x-ray shifts is always greater than 1 meV. If this could be reduced to the level at which the uncertainty in the evaluation of the specific mass shift becomes significant, say 0.2 meV, then the x-ray isotope shifts could be split up into their mass shift and field shift components with two or three times the precision with which this can be done for optical isotope shifts. For example, for ^{202,200}Hg an uncertainty of 0.2 meV in a shift of 150 meV would have to be compared with an uncertainty of 14 MHz in the specific mass shift of an optical transition with a shift of 5300 MHz (see Section 9.80).

The evaluation of $\delta\langle r^2\rangle$ from a field shift in an x-ray transition has been considered by Seltzer (1969). Electron screening effects are simpler than in optical transitions and so more reliable results for $\delta\langle r^2\rangle$ can be obtained. Since the experimental uncertainty in an x-ray shift is quite large, this advantage of x-ray over optical shifts is best utilized by combining the two types of shift in a joint analysis. Consider mercury as an example. For ^{202,200}Hg the value obtained for $\delta\langle r^2 \rangle$ from optical shifts in $\lambda 253.7$ nm is 0.114(8) fm^2. As explained in Section 9.80, the uncertainty arises mainly from the uncertainty in the screening factor used, not from experimental errors. The x-ray field shift in the $K\alpha_1$ line leads to $\delta\langle r^2 \rangle = 0.102(8)$ fm^2, the uncertainty arising mainly from the experimental errors in the measurement of the isotope shift. Shifts have been measured between several isotopes for both x-ray and optical transitions, and the very accurate values of the relative field shifts in the optical transition can be used to obtain an improved set of values of $\delta\langle r^2 \rangle$ from the x-ray data. The technique is illustrated in Fig. 5.1 where the $\delta\langle r^2 \rangle$ values from x-ray shifts are plotted against the optical field shifts. The specific mass shift in the optical transition has been taken as zero for the reasons given in Section 9.80. The best fit line leads to an improved value for $\delta\langle r^2 \rangle$ from the x-ray shifts of 0.106(6) fm^2.

So far it has been assumed that field shifts are proportional to changes in the mean square radius of the nuclear charge distribution, $\delta\langle r^2 \rangle$. Seltzer (1969) showed that this approximation becomes less and less valid the heavier the element under consideration. Moreover, this applies not only to electronic x-ray shifts but to optical shifts as well. The field shift of a single electron was

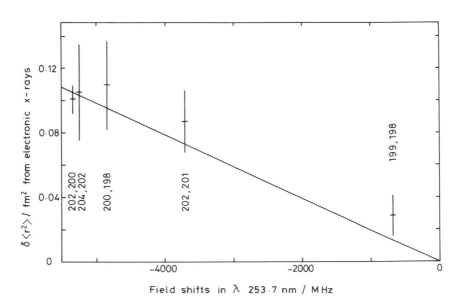

FIGURE 5.1. $\delta\langle r^2 \rangle$ from electronic x-ray shifts plotted against optical field shifts in mercury. The $\delta\langle r^2 \rangle$ values of Lee *et al.* (1978) are plotted against the optical shifts in $\lambda 253.7$ nm of Hg I. The isotope shifts are from Bonn *et al.* (1976); see Section 9.80 for the estimation of zero for the specific mass shift.

given in Eq. (4.24) and can be written

$$\delta E = \pi |\psi(0)|^2 \frac{a_0^3}{Z} f(Z) \delta\langle r^2 \rangle = F\delta\langle r^2 \rangle \tag{5.1}$$

where F is an electronic and not a nuclear factor. This equation was an approximation from Eq. (4.21) and Seltzer gave the better approximation

$$\delta E = F\lambda \tag{5.2}$$

where

$$\lambda = \delta\langle r^2 \rangle + \frac{G}{F}\delta\langle r^4 \rangle + \frac{H}{F}\delta\langle r^6 \rangle + \cdots \tag{5.3}$$

He gave values for the quantities F, G, and H, the ratios of which are very similar for any configuration of an element but vary with the atomic number of the element. In fact, Seltzer (1969) used the notation

$$\lambda = \delta\langle r^2 \rangle + \frac{C_2}{C_1}\delta\langle r^4 \rangle + \frac{C_3}{C_1}\delta\langle r^6 \rangle + \cdots \tag{5.4}$$

but it should be noted that these C's are not related to the isotope shift constant (C), traditionally used in work on optical isotope shifts as defined by Eq. (4.25). The relationship between $\delta\langle r^2 \rangle$ and λ can only be evaluated if the nuclear charge distributions are known in some detail. Using simple models of spherical nuclei it has been calculated that $\lambda = 0.96\,\delta\langle r^2 \rangle$ for $Z = 60$ (Boehm and Lee, 1974) and that $\lambda = 0.93(2)\,\delta\langle r^2 \rangle$ for $Z = 82$ (Thompson *et al.*, 1983).

Most of the isotope shifts measured in electronic x-ray $K\alpha$ lines are given by Boehm and Lee (1974). More recent measurements have been made in mercury (Lee *et al.*, 1978) and cadmium (van Eijk *et al.*, 1979).

5.2. MUONIC ATOMS

In electronic atoms the isotope shift depends on the electron wavefunction at the nucleus and to a good approximation this does not vary across the nuclear volume. Because of this, the field shift is proportional to $\delta\langle r^2 \rangle$ to a good approximation. The muon, being over two hundred times as massive as an electron, sits much closer to the nucleus and its wavefunction varies significantly across the nuclear volume. This is shown schematically in Fig. 5.2 where it is obvious that muons in different levels will probe different properties of the nuclear charge distribution, and the field shifts will not, in general, be proportional to $\delta\langle r^2 \rangle$. For example, a $2s_{1/2}$ muon will hardly interact with the nuclear charge near the surface, whereas a $2p_{3/2}$ muon will do so predominantly. A muonic atom consists of electrons as well as the muon. Soon after capture, while the muon is still in very excited levels, the muon interaction with electrons is complicated; when the muon has reached the lower levels where it interacts strongly with the nucleus, it lies inside all the electrons and can be described very accurately as interacting only with the nucleus. It can be treated as a Dirac particle in the electrostatic field of the nucleus to which various quantum electrodynamic corrections have to be made before comparing its energy with that measured experimentally. The theory of a muonic atom is thus analogous to that of one-electron ions. In both cases it is possible to make an *ab initio* evaluation of the energy of a level for a hypothetical point nucleus and hence, in the muonic case, to determine something about the size of the nuclear charge distribution of an individual nuclide as well as something about the difference between isotopes.

The obvious question is, what property of the nuclear charge distribution is probed by the muon? Figure 5.2 is for a heavy nucleus, and in this case the muon level which has the largest field shift, the $1s_{1/2}$ level, has a wavefunction which falls off significantly between the center and the edge of the nucleus. But for a lighter element there will be two changes: the nuclear radius will be smaller by a factor of roughly $A^{1/3}$, and the size of the muon orbit will be larger by a factor of roughly Z as it is like a one-electron orbit. It follows that, for light elements, the $1s_{1/2}$ muon wavefunction is almost constant across the nuclear volume and so, as in electronic atoms, the field shift is proportional to $\delta\langle r^2 \rangle$ and the energy shift from a hypothetical point nucleus gives a measure of $\langle r^2 \rangle$, the mean square radius of the nuclear charge distribution.

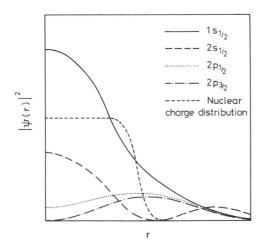

FIGURE 5.2. A schematic diagram (not to scale) of various muonic wavefunctions relative to the nuclear charge distribution of a heavy nucleus.

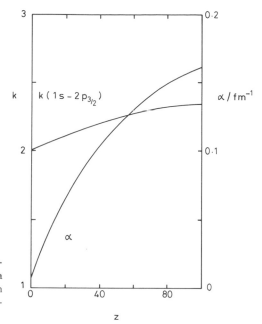

FIGURE 5.3. Parameters of the generalized moment $\langle r^k \exp(-\alpha r) \rangle$ for a Fermi distribution with $c = 1.12 A^{1/3}$ fm and $a = 0.5$ fm as determined by Barrett (1970).

So for light elements, optical, electronic x-ray, and muonic x-ray, field shifts are all probing the same property of the nuclear charge and are directly comparable. For heavy elements this is not so, and a direct comparison of muonic and electronic field shifts can give misleading results as discussed in Section 6.2. Barratt (1970) showed that the muonic energies were best interpreted in terms of the generalized moments $\langle r^k \exp(-\alpha r) \rangle$ of the nuclear charge distribution where k and α are smoothly varying functions of Z for a transition between given muonic levels and α is almost the same for all transitions in a particular element.

Such Barrett moments have been found to be almost independent of what assumptions are made about the actual shape of the nuclear charge distribution and are so said to be "model-independent" (Barrett, 1974). On the assumption of a Fermi distribution, Fig. 4.1b, Barrett obtained the values of k and α shown in Fig. 5.3. For light elements, k is about 2 and α is about 0 so the generalized moment reduces to $\langle r^2 \rangle$, but for heavy elements it differs significantly from $\langle r^2 \rangle$. Because of this, muonic results are usually now expressed in terms of the radius of a nucleus of uniform charge density that has the same moment $\langle r^k \exp(-\alpha r) \rangle$ as the actual nucleus. This equivalent uniform radius, R_k, is defined by

$$\langle r^k \exp(-\alpha r) \rangle = 3 \int_0^{R_k} r^{k+2} \exp(-\alpha r) \, dr / (R_k^3 / 3) \qquad (5.5)$$

TABLE 5.1

Muonic Atom Binding Energies (keV) and Nuclear Charge Radii (fm) of Iron[a]

	$1s_{1/2}$ ^{56}Fe	$2p_{3/2}$ ^{56}Fe	$1s_{1/2}-2p_{3/2}$		
			^{56}Fe	^{54}Fe	^{56,54}Fe
Experimental energy	1734.267	477.213	1257.054	1260.011	−2.957
Corrections					
Nuclear polarization	0.589	0.007	0.582	0.546	0.036
Mass shift	0.058	0.003	0.055	0.057	−0.002
QED, etc.	11.487	1.824	9.663	9.711	−0.048
Uncorrected energy	1722.133	475.379	1246.754	1249.697	−2.943
c (of Fermi distribution)			4.04735	3.98097	
R_k ($k = 2.121$)			4.7941	4.7387	0.0554
$\langle r^2 \rangle^{1/2}$			3.743	3.700	0.043

[a] Figures from Shera *et al.* (1976) who fitted the uncorrected energy (energy of a Dirac particle in the field of an infinitely heavy nucleus) to a Fermi charge distribution with $a = 0.55$ fm; $\alpha = 0.074$ fm^{-1} was used in the Barrett moment.

This is a generalization of Eq. (4.3), the latter being the particular case when $k = 2$ and $\alpha = 0$. The main advantage of quoting R_k rather than $\langle r^2 \rangle$ is that it is much more model-independent. For different muonic transitions, k can vary from about 1 to about 4, but the transition $1s-2p_{3/2}$, with k equal to a little more than 2, is the one for which the nearest to a direct comparison with electronic field shifts can be made. The best comparison is made by using as many values of k as possible, even though none of them is equal to 2. If many Barrett moments are known, then $\delta(\langle r^2 \rangle^{1/2})$ can be determined in a more and more model-independent way as more and more different Barrett moments are considered. To obtain more different moments, electron scattering results can be considered in addition. Wohlfahrt *et al.* (1980) have made such a combined analysis to obtain values of $\delta(\langle r^2 \rangle^{1/2})$ which may be compared directly with results from electronic field shifts.

Some results for iron, a fairly light element, taken from Shera *et al.* (1976) are given in Table 5.1. It can be seen that some large corrections have to be subtracted from the experimental energies to arrive at the uncorrected energy. This is the energy of a Dirac particle in the static Coulomb field of the nucleus with the charge distribution of the real nucleus but of infinite mass. Comparison of this energy with the value calculated *ab initio* for a point nucleus gives the size of the nuclear charge distribution. It can also be seen from Table 5.1 that the advantage of considering isotope shifts between isotopes is that these corrections tend to cancel out.

The nuclear polarization correction allows for the virtual excitation of the nucleus from its ground state by the muon. It is necessary to have good information about the properties of the low-lying excited states of the nuclide before this correction can be calculated. Even if the nucleus is treated as being in its ground state with a static charge density, there are various quantum electrodynamic corrections to make. The dominant correction is for the vacuum polarization due to $\mu^+\mu^-$ pairs. The mass shift is relatively much smaller than it is in optical isotope shifts, and although in principle it is not so simple as for one-electron atoms because relativistic effects must be included (Barrett *et al.*, 1973), in practice these difficulties can usually be ignored as the effects are not as large as the uncertainty in the nuclear polarization correction. This uncertainty is also very much reduced when differences between isotopes are considered. For the figures given in Table 5.1, the uncertainty in the nuclear polarization is about 0.23 keV for each isotope but only about 0.04 keV for the isotope shift. The experimental (total) energies are also uncertain by about 0.04 keV which leads to an uncertainty of about 0.0008 fm in R_k and of about 0.0006 fm in δR_k. The values for $\langle r^2 \rangle^{1/2}$ are rather model-dependent, but Wohlfahrt *et al.* (1980), by considering electron scattering data in addition, have deduced that $\delta(\langle r^2 \rangle^{1/2}) = 0.0445(8)$ fm for ^{56,54}Fe.

Much else can be deduced from muonic x-ray energy measurements; this section has dealt only with those features that are relevant when comparisons are being made with isotope shifts in electronic atoms. The subject is dealt with much more fully and authoritatively by Barrett and Jackson (1977). Information about nuclear sizes other than isotope shifts which can be obtained from muonic atoms will be considered further in Section 7.2.

SEPARATION OF MASS AND FIELD SHIFTS IN OPTICAL ISOTOPE SHIFTS

6.1. SEPARATION USING OPTICAL ISOTOPE SHIFTS ONLY

The isotope shift is to a very good approximation the sum of the mass shift and the field shift

$$IS = MS + FS \tag{6.1}$$

Eq. (3.13) is, when generalized to the many-electron case,

$$MS_{\infty, M} = \frac{\langle \Sigma_i p_i^2 \rangle}{2(M+m)} + \frac{\langle \Sigma_{i>j} \mathbf{p}_i \cdot \mathbf{p}_j \rangle}{M+m} \tag{6.2}$$

where $MS_{\infty, M}$ is the mass shift between isotopes with nuclear masses ∞ and M. Eq. (6.2) can be written as

$$MS_{\infty, M} = \frac{K}{M+m} \tag{6.3}$$

where K is a constant that is independent of the mass M. For a shift between two isotopes with nuclear masses M_P and M_Q

$$MS_{P,Q} = \frac{K(M_P - M_Q)}{(M_P + m)(M_Q + m)} \tag{6.4}$$

If the isotope shifts $IS_{P,Q}$ in a spectral line are multiplied by a modifying factor

$$\mu_{(P,Q)} = \frac{(M_S - M_T)(M_P + m)(M_Q + m)}{(M_P - M_Q)(M_S + m)(M_T + m)} \tag{6.5}$$

then the modified mass shifts will be equal for all the pairs of isotopes; the modified mass shift is $MS_{S,T}$, the isotope pair M_S, M_T being treated as the standard pair of isotopes.

Suppose that isotope shifts have been measured between various isotopes P, Q, R, S, T, etc., in two spectral lines a and b. The modified shifts are

$$\mu_{(P,Q)} IS_{P,Q}^a = MS_{S,T}^a + \mu_{(P,Q)} FS_{P,Q}^a \tag{6.6}$$

and

$$\mu_{(P,Q)} IS^b_{P,Q} = MS^b_{S,T} + \mu_{(P,Q)} FS^b_{P,Q} \qquad (6.7)$$

A plot of the modified shifts in spectral line *b* against the modified shifts in spectral line *a* should give a straight line as shown in Fig. 6.1. This line passes through the point $M^a_{S,T}$, $M^b_{S,T}$, and has a gradient FS^b/FS^a. The latter can be determined from the plot; unfortunately, the former cannot. This relationship between isotope shifts was pointed out by King (1963) and plots like Fig. 6.1 are often called King plots.

A simpler modifying factor that is sometimes used is

$$\mu' = \frac{(M_P + m)(M_Q + m)}{M_P - M_Q} \qquad (6.8)$$

This gives a line of the same gradient but passing through the point K^a, K^b, K being defined as in Eq. (6.3). This technique has the disadvantage that the modified shifts are not of similar size to the size of the measured shifts. It has

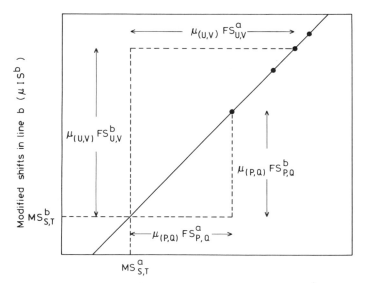

FIGURE 6.1. A King plot. The modified shifts in spectral line b are plotted against the modified shifts in spectral line a. The results for four pairs of isotopes are shown as (●). The division between mass shifts and field shifts is shown for two of the pairs.

the advantage that K is independent of the masses of the isotopes under consideration, and so is more relevant when comparing one element with another.

The modified shifts in other spectral lines c, d, and so on, can be plotted on the same diagram. This is done for isotope shift measurements made in samarium by Striganov *et al.* (1962) in Fig. 6.2. In order to reduce the error in the abscissa, the modified isotope shifts of the various spectral lines are plotted against the mean of the modified isotope shifts in a few of the lines in which similar shifts were measured with the smallest uncertainty. Since the plot involves linear relationships, it works for linear combinations of shifts in different spectral lines. The plot in Fig. 6.2 concentrates on the region of particular interest, which is the region of the mass shifts. The experimental points are off to the left, the values of the modified shifts on the abscissa, range from -1200 MHz for ^{154,152}Sm to -2180 MHz for ^{152,150}Sm. There is, of course, some uncertainty in the extrapolation into Fig. 6.2 and the position of each line is uncertain by about 100 MHz. In 1963 the sizes of the mass shifts were not known, it was assumed that they were in the region of zero to about three times the normal mass shift. The various spectral lines are identified by their wavelength in nm in Fig. 6.2 and the normal mass shift for ^{154,152}Sm (the standard pair of isotopes) is about 30 MHz. Although Fig. 6.2 does not give the mass shifts, it does give the set of mass shifts in all the spectral lines if the mass shift in one spectral line is arbitrarily (or for some good reason) chosen to have a particular value. For instance, if the mass shifts in $\lambda 497.6$ nm and $\lambda 525.2$ nm are set at zero then the mass shifts in the other spectral lines are determined by where their plotted lines in Fig. 6.2 cross the vertical line A. In this case the other mass shifts range from 135 MHz in $\lambda 532.1$ nm up to 570 MHz in $\lambda 587.1$ nm. The latter is about twenty times the normal mass shift. Wherever the vertical line is drawn in Fig. 6.2 it gives some mass shifts that are many times the normal mass shift. The position of the vertical line which gives the smallest maximum mass shift is shown as B. In this case the mass shifts range from 45 MHz in $\lambda 621.8$ nm to 465 MHz in $\lambda 587.1$ nm and $\lambda 532.1$ nm. The latter is about fifteen times the normal mass shift. Even allowing for the uncertainty in the extrapolation procedure, the inevitable (and in 1962, surprising) conclusion is that some of the mass shifts are at least ten times as big as the normal mass shift. It can also be deduced from Fig. 6.2 that the large mass shifts are probably negative in the transitions studied. It was Bauche (1969) who showed by Hartree–Fock calculations that these large negative mass shifts arise in transitions of the type $4f^n - 4f^{n-1}$, i.e., where the lower level has an extra 4f electron. The lines with positive isotope shifts, that is positive field shifts, are of the type $4f^6 6s^2 - 4f^5 5d6s^2$, and those with negative shifts of the type $4f^6 6s^2 - 4f^6 6s6p$. Bauche calculated that the specific mass shift in a $4f^6 - 4f^5$ transition would be 1000 MHz and the vertical line C in Fig. 6.2 is in approximate agreement with this and the configurations involved in the transitions. This technique was used to point out the existence of large and hitherto unexpected mass shifts in samarium (King, 1963) and has since been applied to much more accurate isotope shift data. The isotope shifts between even isotopes lie almost on straight lines in all cases except one. That one is samarium (Griffith *et al.*, 1979), and is discussed in Section 9.62. If odd isotopes are involved, then it is often good enough to

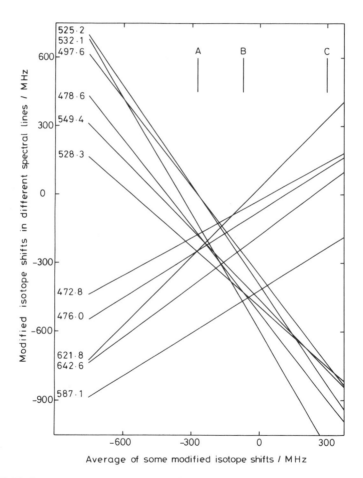

FIGURE 6.2. Isotope shifts in samarium. The shifts in various spectral lines, modified to give the mass shift for ^{154,152}Sm, are plotted against the average of the modified shifts in a number of spectral lines along the abscissa. The spectral lines are identified by their wavelengths in nm on the left. The three vertical lines, A, B, and C, give three sets of mass shifts which are discussed in the text. The isotope shifts were measured by Striganov *et al.* (1962).

work with the center of gravity of the magnetic hyperfine structure components, but in very accurate work, allowance must be made for hyperfine perturbation of one fine structure level by another level. This is discussed in Section 9.3.1, where the hyperfine perturbation affects the mass shifts in Li II, and in Section 9.80 where it affects the field shifts in Hg I. The perturbation is often referred to as the second-order hyperfine interaction.

Another example of the use of a plot between optical isotope shifts is in the spectra of barium. Isotope shifts have been measured between most of the isotopes between ^{138}Ba and ^{126}Ba in the line of Ba I with a wavelength of 553.5 nm. These measurements have been made with great accuracy using Doppler-free laser spectroscopy techniques (Baird *et al.*, 1979; Bekk *et al.*, 1979). Some earlier measurements in the line of Ba II with a wavelength of 493.4 nm were made by Fischer *et al.* (1974a). Their accuracy is relatively poor and they studied fewer isotopes (although ^{140}Ba is unique to this work) but they were able to calculate the size of the mass shift as -9 MHz for ^{138,136}Ba. This Hartree–Fock evaluation should be reasonably reliable as the transition (an alkali-like resonance line) is about the simplest possible. It is difficult to estimate the uncertainty, but it is probably about 12 MHz. A plot of the isotope shifts in $\lambda 553.5$ nm against those in $\lambda 493.4$ nm allows the mass

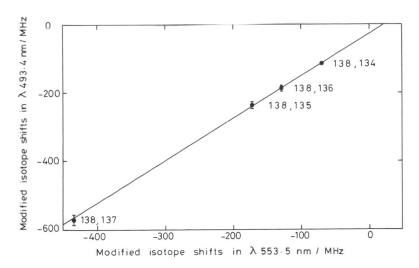

FIGURE 6.3. Optical isotope shifts in barium. The shifts, modified to give the mass shift for ^{138,136}Ba, in $\lambda 493.4$ nm of Ba II ($6s\,^2S_{1/2} - 6p\,^2P_{1/2}$) are plotted against the modified shifts in $\lambda 553.5$ nm of Ba I ($6s^2\,^1S_0 - 6s6p\,^1P_1$). The equation of the line of best fit is $y = 1.252(7)x - 24(1)$. The uncertainties in the shifts in $\lambda 553.5$ nm are too small to show in the diagram, being 1 MHz or less. The isotope pairs are identified by their mass numbers. The shifts in $\lambda 493.4$ nm were measured by Fischer *et al.* (1974a) and those in $\lambda 553.5$ nm by Baird *et al.* (1979) and Bekk *et al.* (1979).

shift and the isotope shift of ^{140}Ba to be found for the former line. The plot is shown in Fig. 6.3. It can be deduced from this plot that in λ553.5 nm, the mass shift is 12(12) MHz for ^{138,136}Ba and that the isotope shift (not modified) for ^{140,138}Ba is −923(100) MHz. The case of the λ553.5 nm line of barium will be returned to in Section 6.2.

It sometimes happens that in a spectrum which in general exhibits field shifts, there is a line which has mass shifts but no field shifts or very small field shifts. A King plot of such a line will reveal this fact so its mass shift can be determined. Unfortunately, in this case, knowing the mass shift in one spectral line does not allow the mass shifts in the other spectral lines to be determined from the plot. An example of this arises in cerium and is shown in Fig. 6.4.

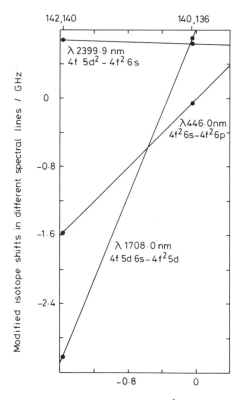

FIGURE 6.4. Optical isotope shifts in Ce II. The shifts have been modified to give the mass shift for ^{142,140}Ce. Data are from Champeau and Verges (1976). The spectral line of wavelength 2399.9 nm has a mass shift of about 0.65 GHz. Assuming the other transition involving 4f–4f^2 has a similar mass shift, then the mass shift in λ446.0 nm is relatively small, as would be expected in a 6s–6p transition.

There is also the possibility that a spectral line could have field shifts but no mass shift. If the modified shifts in such a spectral line are plotted against those in another spectral line, the lack of a mass shift is not revealed.

6.2. COMBINING OPTICAL ISOTOPE SHIFTS WITH OTHER DATA

As already explained in Chapter 5, it is possible to calculate the size of the mass shifts in x-ray isotope shifts of electronic atoms. These can be subtracted to give the field shifts in x-ray transitions which can be plotted against isotope shifts in optical transitions to find the mass shifts in the latter. For example, the x-ray isotope shifts in mercury (Lee *et al.*, 1978) show that the mass shift for ^{202,200}Hg in λ253.7 nm is 1.1(1.4) GHz. The rather large uncertainty arises from the scatter of the points when the modified x-ray shifts in the $K\alpha_1$ line are plotted against the modified optical shifts in λ253.7 nm ($6s^2\,{}^1S_0 - 6s6p\,{}^3P_1$). A mass shift of 1 GHz corresponds to a Vinti k factor of 6 a.u., so it is extremely unlikely that the mass shift is anything like as large as this in a 6s–6p transition.

A more accurate estimate of the mass shift in an optical transition can usually be made by plotting against a muonic x-ray transition, rather than an electronic x-ray transition (assuming that the muonic shifts have been measured between suitable pairs of isotopes). As an example, let us return to the case of barium. As already mentioned in Section 6.1, the isotope shifts between many isotopes have been measured very accurately in λ553.5 nm of Ba I. A comparison with λ493.4 nm of Ba II shows that the mass shift in λ553.5 nm is 12(12) MHz for ^{138,136}Ba.

Muonic x-ray transition energies have been measured for several isotopes of barium (Shera *et al.*, 1982) and modified isotope shifts derived from these can be plotted against modified shifts in λ553.5 nm of Ba I. This has been done in Fig. 6.5, and the resultant best-fit plot is a very plausible straight line. The mass shift in the muonic $K\alpha_1$ line is 0.05 keV, and so it can be deduced from Fig. 6.5 that the mass shift in λ553.5 nm of Ba I is 112(13) MHz for ^{138,136}Ba. This result is at variance with that already obtained by a consideration of optical isotope shifts alone, namely 12(12) MHz.

The result obtained from Fig. 6.2 is wrong because the muonic isotope shifts do not depend on the nuclear charge distributions in the same way as do optical isotope shifts. Although Fig. 6.5 looks plausible it is not sensible. A muonic K x-ray energy depends on a property of the nuclear charge distribution of that isotope; however, differences of this energy between isotopes do not depend on $\delta\langle r^2\rangle$ but on $\delta\langle r^k\exp(-\alpha r)\rangle$ as explained in Section 5.2. Shera *et al.* (1982) measured the energies of other muonic x-ray transitions, which involved different values of k and so gave various properties of the nuclear charge distribution. From these various properties they deduced

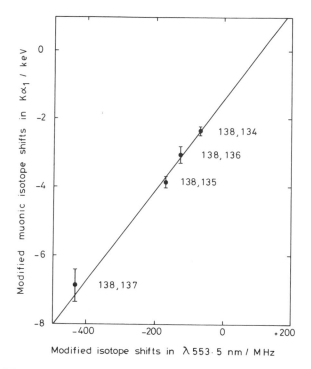

FIGURE 6.5. Muonic isotope shifts plotted against optical isotope shifts in barium—a spurious King plot. The shifts, modified to give the mass shift for ^{138,136}Ba, in the $K\alpha_1$ muonic transition are plotted against the modified shifts in λ553.5 nm of Ba I. The uncertainties in the latter are 1 MHz or less. The muonic shifts were measured by Shera *et al.* (1982), the optical shifts by Baird *et al.* (1979) and Bekk *et al.* (1979). Although the data lie on a straight line, the predicted mass shift of 112(13) MHz in λ553.5 nm (based on a mass shift of 0.05 keV in the muonic transitions) is erroneous, as explained in the text.

values of $\langle r^2 \rangle$ (see Fig. 2.1 in Section 2.4). The differences in these between isotopes ($\delta\langle r^2 \rangle$), suitably modified, are plotted against the modified optical shifts in λ553.5 nm in Fig. 6.6. The mass shift in λ553.5 nm is given by the value of the plotted line at $\delta\langle r^2 \rangle = 0$. The value obtained from the best fit line of Fig. 6.6 is 7(11) MHz which is in excellent agreement with the value of 12(12) MHz already obtained. Combining the two results, the mass shift in λ553.5 nm of Ba I is 9(9) MHz for ^{138,136}Ba. The determination of $\delta\langle r^2 \rangle$ for barium isotopes using this result has already been discussed briefly in Section 2.4 and is considered in more detail in Section 9.56.

Muonic isotope shifts have often been plotted against optical isotope shifts to find the mass shifts in the latter. Although this technique is not strictly valid, as explained above and, as pointed out by Stacey (1971), the above

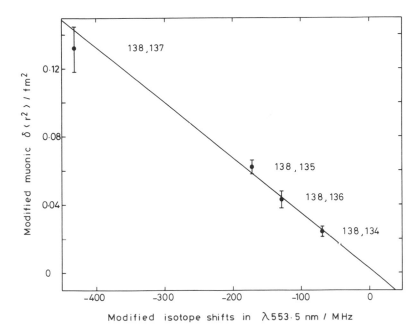

FIGURE 6.6. $\delta \langle r^2 \rangle$ from muonic data plotted against optical isotope shifts in barium. The $\delta \langle r^2 \rangle$ values of Shera *et al.* (1982) are plotted against the optical shifts in $\lambda 553.5$ nm of Ba I $(6s^2 {}^1S_0 - 6s6p {}^1P_1)$. The equation of the line of best fit is $y = -0.00033(3)x + 0.002(4)$. The uncertainties in the shifts in $\lambda 553.5$ nm are too small to show in the diagram, being 1 MHz or less. The optical data is from Baird *et al.* (1979) and Bekk *et al.* (1979). All results have been modified to give the mass shift for ^{138,136}Ba. The predicted mass shift in $\lambda 553.5$ nm is 7(11) MHz.

example of barium shows that the error in so doing is not too large in some circumstances. The difference between the two mass shift estimates of 112(13) MHz, and 12(12) MHz only appears to be large because both the muonic and the optical isotope shifts have been measured with great precision. In much earlier work an error of 100 MHz in the estimation of a mass shift was not significant.

RELATED TECHNIQUES FOR THE DETERMINATION OF NUCLEAR STRUCTURE

7.1. ELASTIC ELECTRON SCATTERING

The elastic scattering of electrons by nuclei is caused by the electrostatic field of the nucleus, and so the analysis of such scattering experiments gives information about nuclear charge distributions. Isotope shifts can, in principle, give values of the change between isotopes in the mean square radius of the nuclear charge distribution, $\delta\langle r^2 \rangle$, but sometimes in practice they give only relative values between different isotope pairs. In such cases, inaccurate (but absolute) values of $\delta\langle r^2 \rangle$ from other techniques can be combined with accurate (but only relative) values of $\delta\langle r^2 \rangle$ from optical isotope shift work. Iron is an example of where this technique might be used, but the information available from electron scattering is of little help; $\langle r^2 \rangle$ has been determined for ^{54}Fe, ^{56}Fe, and ^{58}Fe, but the uncertainties in the values are about a quarter of their differences between isotopes. Muonic x-ray energy studies on the other hand can give values of $\delta\langle r^2 \rangle$ of a useful precision for combination with optical isotope shift measurements. This often seems to be the case but, even so, the importance of electron scattering data should not be underestimated. The analysis of muonic x-ray energies to give precise information about nuclear charge distributions would be impossible without the information that had previously been obtained from electron scattering experiments about nuclear charge distributions. Since the first measurements of Lyman *et al.* (1951), the analysis of electron scattering experiments has given more information on nuclear charge distributions than has any other technique. From the scattering at one electron energy it is not possible to deduce the actual nuclear charge distribution, but only a set of charge distributions that are consistent with the scattering data. However, by repeating the experiment with different electron energies it is possible to obtain a fairly unique distribution from what is common to the different sets of charge distributions.

It was gradually discovered that the two-parameter Fermi distribution gave a good description of almost all nuclear charge distributions (see Fig. 4.1b). At one time it was thought that there might be oscillations in the nuclear charge density, $\rho(r)$, with radius r, and in particular that the charge density might be quite low near the center of the nucleus. Such "bubble" nuclei were consistent with low-energy electron scattering data and 1s muonic energies, but it was shown that distributions with increased density at the center were also consistent. The use of the model-independent Barrett moments, which have already been mentioned in connection with isotope shifts in muonic atoms in Section 5.2, enabled the issue to be resolved. For further information, see the reviews by Friar and Negele (1975) and Barrett and Jackson (1977).

So far as comparison with isotope shift work is concerned, where the relevant Barrett moment k is two, low-energy electron scattering results, where k is also two, can be compared directly. But as already pointed out, the main importance of electron scattering work is the overall picture it has given of nuclear charge distributions; within this overall picture muonic x-ray and optical isotope shifts can highlight some of the details. In conclusion, it should be pointed out that the most detailed information about the nuclear structure of the lightest elements is obtained from electron scattering data. Such detailed information can be used to determine the amount of nuclear polarization, the biggest unknown, in muonic energy levels of light elements (see Sick, 1982).

7.2. MUONIC ATOM ENERGIES

Muonic atom energies have already been mentioned in Section 5.2, where the stress was laid on the isotope shift in muonic levels and the transitions between levels. In the case of muonic levels it is possible to deduce something about the size of the nuclear charge distribution for a particular nuclide as well as about the difference in size between isotopes. This is because the effects of the electrons on the muon can be treated as small perturbations for low-lying muon levels and it is just these low-lying ones that are of interest because of their strong nuclear interaction. The muon–nucleus system can be handled by an *ab initio* theory. The muon is treated as a Dirac particle in the Coulomb field of an infinitely heavy nucleus of finite size. Before this can be compared with experimental results, various corrections have to be made, as described in Section 5.2. The biggest uncertainty that arises in these corrections is in the determination of the nuclear polarization. Having made these corrections to the experimentally measured energy, a fit to the theoretical value is obtained by adjusting the nuclear charge distribution. If energies are available for more than one muonic level then, as already explained in Section 5.2, more than one property of the nuclear charge distribution can be determined.

It is usually assumed that the nuclear charge follows a two-parameter Fermi distribution described by c and t, as shown in Fig. 4.1b. One muonic energy level can be interpreted as various combinations of c and t, it does not give a unique pair of values any more than do the electron scattering data.

However, other muonic levels give other sets of values and by combining them, a unique value for c and t can be obtained. The way in which this can work is shown in Fig. 7.1. The technique was used, for example, by Kessler *et al.* (1975) with their muonic energy measurements in lead. They measured the energies of several levels as well as the usual $1s_{1/2}$ level and all gave about the same value for c and t on a so-called "$c - t$ plot."

Since the 1s–2p transition can be studied with greater precision than any other, the analysis is often based on this transition alone. As mentioned in Section 5.2, the least model-dependent property of the nuclear charge distribution that is obtained from the measurement of muonic energy levels is $\langle r^k e^{-\alpha r} \rangle$, where k and α have been determined for different transitions in different atoms (Barrett, 1970). Results that have been obtained for 1s–2p muonic transitions in various elements are given in Table 7.1. This is just a small sample spanning the Periodic Table; many more nuclear charge distributions deduced from muonic x-ray energies are given in Chapter 9. For the lightest elements the largest uncertainty is in the measurement of the x-ray transition energy, but in the heavier elements it is in the determination of the nuclear polarization. As can be seen, these uncertainties are small compared with the energy difference between a point nucleus and the actual nucleus—the finite-size effect. It follows that the fairly model-independent equivalent uniform radii R_k, defined by Eq. (5.5), can be calculated with considerable accuracy. The values of $\langle r^2 \rangle^{1/2}$, on the other hand, are much more model-dependent, and it is difficult to estimate what uncertainties should be attached to them to make them model-independent. A combined analysis of muonic x-ray energies and electron scattering data can help in this respect; such combined analyses have been made by Wohlfahrt *et al.* (1980) for iron, nickel,

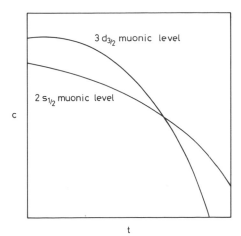

FIGURE 7.1. A $c - t$ plot from muonic energy measurements in two different muonic levels. The point of crossover gives the unique value of the parameters c and t in the Fermi distribution of the nuclear charge.

TABLE 7.1

1s–2p Muonic X-Ray Energies and Nuclear Charge Distributions

	$^9\text{Be}^a$	$\sim ^{39}\text{K}^b$	$^{152}\text{Sm}^c$	$^{208}\text{Pb}^d$
Experimental energy (keV)	33.402	711.85	4,458.6	5,962.88
Nuclear polarization (keV)	0.001	0.12	5.3	5.00
Finite-mass effect (keV)	0.0	0.02	0.2	0.28
Finite-size effect (keV)	−0.074	−56.45	−4,238.8	−10,233.51
k	2.116	2.112	2.277	2.47
α/fm	0.042	0.064	0.125	0.17
R_k/fm	3.07	4.409	6.497	7.033
$\langle r^2 \rangle^{1/2}$/fm	2.39	3.437	5.083	5.510

[a] Schaller *et al.* (1980).
[b] Schaller *et al.* (1978).
[c] Powers *et al.* (1979).
[d] Barrett and Jackson (1977); Ford and Rinker (1973); Kessler *et al.* (1975).

and zinc, and by Euteneuer *et al.* (1978) for bismuth, lead, and thallium, to give but two examples from very different parts of the Periodic Table.

Even though muonic x-ray energies can now be measured to something approaching one part in 10^5, *ab initio* evaluations can match this precision by using very refined and complicated QED calculations. There do not appear to be any significant discrepancies between experiment and theory. For low-lying muonic levels this is not surprising because any discrepancy can be removed by altering the assumed size of the nuclear charge distribution. Since the finite-size effect makes such a large contribution, the required alteration is relatively so small that it does not introduce any anomaly with other measurements of nuclear size. For higher muonic levels the situation is different, however; here, the finite-size effect is very small. The contributions from higher-order QED corrections and the screening of the muon by the atomic electrons (Vogel, 1974) are larger, and must be correctly evaluated if theory and experiment are to agree. Even for these higher levels the agreement is, in general, very reasonable. For further details, see the reviews by Barrett and Jackson (1977) and Borie and Rinker (1982). The information on nuclear charge distributions, which had been obtained at the time from muonic atoms, was collected together and reviewed by Engfer *et al.* (1974).

7.3. ISOMER SHIFTS

7.3.1. ISOMER SHIFTS IN OPTICAL SPECTRA

An isomer is a nuclide in an excited state which is sufficiently long-lived for its properties in that state to be measured. If the mean square radius of the nuclear charge distribution of the isomer differs from that of the nuclide in the ground state, then the frequency of a spectral line will show a field shift between the isomer and the ground state; such a field shift is called an isomer shift. The measured isomer shift is the field shift; there is no mass shift to subtract since the isomer and the ground state nuclide have the same mass. The first measurement of an isomer shift was by Melissinos and Davis (1959) for ^{197}Hg. Isomer shifts have now been measured in six elements and the most recent reference in each case is: rubidium, Thibault *et al.* (1981a); silver, Fischer *et al.* (1975b); caesium, Thibault *et al.* (1981b); barium, Bekk *et al.* (1979); mercury, Dabkiewicz *et al.* (1979); thallium, Goorvitch *et al.* (1969); lead, Thompson *et al.* (1983). The significance of these results is discussed in the relevant sections of Chapter 9 and in Section 10.2.

7.3.2. MUONIC ISOMER SHIFTS

The isomer shift in optical spectra considered in the previous section has its direct equivalent in muonic x-ray spectra, but such shifts are not easily measured in practice. What can be measured more easily is a shift in the nuclear γ-ray transition. Returning, for a moment, to the optical isomer shift, it can crudely be described in terms of the sum of three energies: N, the energy of a bare nucleus, E the energy of the electrons around a point nucleus, and N,E the energy due to the interaction of the electrons with the nuclear charge of finite size, in other words, the field shift relative to a point nucleus. Describing the upper and lower electronic states by E' and E'', and the isomeric and ground nuclear states by N^* and N^0, the optical isomer shift is given by

$$[(N^* + E'' + N^*,E'') - (N^* + E' + N^*,E')]$$

$$- [(N^0 + E'' + N^0,E'') - (N^0 + E' + N^0,E')] \qquad (7.1)$$

$$= N^*,E'' - N^*,E' - N^0,E'' + N^0,E'$$

which can be written

$$= (N^* - N^0),E'' - (N^* - N^0),E' \qquad (7.2)$$

which is just the optical isomer shift in the lower level less that in the upper level. The directly equivalent shift in a muonic x-ray spectrum would be

$$(N^* - N^0),\mu'' - (N^* - N^0),\mu' \qquad (7.3)$$

but, as already mentioned, such shifts are difficult to measure.

When a muonic atom is formed, the muon is initially in a highly excited state. As it cascades down to the lower states discussed in Sections 5.2 and 7.2, there is a possibility of the nucleus being excited, particularly if the nuclear excitation energy is nearly resonant with a muonic transition energy. If such an excited nucleus is not very short-lived then it may not decay to the ground state until the muon is in a 1s state. In this case, the strong interaction of the muon with the nuclear charge distribution will affect the energy of the γ-ray emitted during the nuclear decay. This effect is measured by comparing the γ-ray energy of muonic atoms with that of ordinary atoms, the difference is the muonic isomer shift. Using the above notation, with μ'' for the energy of a 1s muon around a point nucleus, the muonic isomer shift is given by

$$\left[(N^* + \mu'' + N^*,\mu'') - (N^0 + \mu'' + N^0,\mu'') \right]$$

$$- \left[(N^* + E'' + N^*,E'') - (N^0 + E'' + N^0,E'') \right] \qquad (7.4)$$

$$= N^*,\mu'' - N^0,\mu'' - N^*,E'' + N^0,E''$$

which can be written

$$= (N^* - N^0),\mu'' - (N^* - N^0),E'' \qquad (7.5)$$

and since the electronic interaction is negligible compared with the muonic 1s interaction

$$\simeq (N^* - N^0),\mu''. \qquad (7.6)$$

It is this shift that can be measured more easily, and it can be seen that it is the dominant term of Eq. (7.3).

There are a few complications that must be allowed for [see Barrett and Jackson (1977) and Anigstein et al. (1980)], but there is one simplification compared with muonic isotope shifts. The nuclear polarizations are similar in the ground and excited nuclear states; they can often be considered to be the same and are so eliminated from the isomer shift. The first measurement of a muonic isomer shift was made by Bernow et al. (1967) for ^{152}Sm. Several results were reviewed by Wu and Wilets (1969) and Barrett and Jackson (1977). More recent measurements have been made on ^{150}Sm (Yamazaki et al., 1978), ^{204}Pb (Hoehn et al., 1980), ^{206}Pb (Hoehn et al., 1982), and ^{207}Pb (Anigstein et al., 1980). Yamazaki et al. (1978) also measured the isomer shift of ^{152}Sm in the muonic x-ray spectrum [see Eq. (7.3)], rather than the nuclear γ-ray spectrum [Eq. (7.6)]. Such measurements have also been made in ^{172}Yb

(Hoehn and Shera, 1979) and [186,188,190,192]Os (Hoehn et al., 1981). The isomer shifts of the first excited states of [152,154]Sm have been determined by Powers et al. (1979) from the 2p muonic hyperfine structure. This is possible because when the muon reaches the 2p level, the probability of the nucleus being in the first excited state is nearly 30% and so the relative splittings of the 2p–1s hyperfine structure are affected by the size of the nucleus in its excited state. These isomer shifts are discussed in the relevant sections of Chapter 9 and in Section 10.2.

7.3.3. MÖSSBAUER ISOMER SHIFTS

Just as the muonic isomer shift occurs in both muonic x-ray transitions and nuclear γ-ray transitions, so the electronic isomer shifts of Eq. (7.2) can occur in the nuclear γ-ray transitions

$$
\begin{aligned}
(N^* - N^0), & E'' - (N^* - N^0), E' \\
&= \left[(N^* + E'' + N^*, E'') - (N^0 + E'' - N^0, E'') \right] \\
&\quad - \left[(N^* + E' + N^*, E') - (N^0 + E' + N^0, E') \right]
\end{aligned} \tag{7.7}
$$

The difficulty in measuring such shifts is that the shift as a fraction of the γ-ray transition energy is so minute. A typical isomer shift is 1 GHz, or 4 μeV, and a typical transition energy is 100 keV, so the energy must be measured to at least about one part in 10^{12} to obtain useful information about the shift. This fantastic precision is possible in Mössbauer transitions where there is negligible nuclear recoil, and hence the linewidth of the γ transition is little greater than the natural width.

In Mössbauer spectroscopy a resonance is observed when the energy of nuclear γ-rays emitted from a source material is equal to the energy difference between the nuclear ground state and the excited state in the absorber plus an energy shift introduced by the Doppler effect of a relative motion of the source and absorber. Thus, in Eq. (7.7), E'' is the electronic energy in the source and E' that in the absorber. Their interactions with the nuclear charge are different because the source and absorber consist of two chemically different materials. For example, in Mössbauer's original experiment the source was the β^- emitter [191]Os, which decays through the 129 keV level of [191]Ir; the absorber was [191]Ir, and resonance was observed in the 129-keV γ-rays. Most Mössbauer spectroscopy is devoted to finding out about the changes in N, E, that is, the changes in $|\psi(0)|^2$, when the solid state conditions in the emitter and source are changed. But in those cases where the electronic conditions are known, then information can be obtained from N, E about the change between the isomeric and ground states in the mean square radius of the nuclear charge distribution. Mössbauer isomer shifts involve the same property of the nuclear charge distribution $\delta\langle r^2\rangle$ as do optical isomer shifts, because in both cases the nucleus is interacting with electrons whose wavefunctions are constant over the nuclear volume. On the other hand, the muonic isomer shift involves $\delta\langle r^k\exp(-\alpha r)\rangle$, where k is the moment appropriate to the 1s muonic level, k and α being as defined in Section 5.2.

The results obtained for $\delta\langle r^2\rangle$ in nearly fifty elements ranging from potassium to americium are given by Shenoy and Dunlop in Appendix I of Shenoy and Wagner (1978). The latter covers all aspects of the subject of Mössbauer isomer shifts very thoroughly.

7.4. MEASUREMENT OF NUCLEAR DEFORMATION

7.4.1. COULOMB EXCITATION

Coulomb excitation is an inelastic scattering process in which a charged bombarding particle, such as an α-particle, excites the target nucleus via the electromagnetic field only. Electric quadrupole interaction, in which, for example, a nucleus which initially has $I=0$ is excited to a state with $I=2$, depends on the intrinsic quadrupole moment of the nucleus in its ground state, as in Eq. (4.43). As mentioned in Section 4.2.3, this intrinsic quadrupole moment is related to the quadrupole deformation parameter β_2. The relationship is strictly a model-dependent one, that is to say, it depends on assumptions made about the nuclear charge distribution, but nowadays enough is known about these distributions for the β_2 values obtained from Coulomb excitation work to be quite reliable enough for the determination of the amount of a field shift which arises from the change in deformation between isotopes. The deformations are particularly large where both the number of protons and the number of neutrons in a nuclide lie well away from magic numbers. This happens for reasonably stable nuclei near $Z=35$, $N=40$; $Z=70$, $N=100$; and $Z=90$, $N=140$. These very deformed nuclei show prominent rotational levels and not only the probability of transitions between these levels, but also their energy separation, can give a measure of the nuclear deformation parameter β_2.

7.4.2. SPECTROSCOPIC HYPERFINE STRUCTURE

The dominant effect is the magnetic dipole interaction of the atomic electrons with the magnetic dipole moment of the nucleus. Superimposed on this is the electric quadrupole interaction involving the gradient of the electric field of the electrons at the nucleus and the electric quadrupole moment of the nucleus. This quadrupole interaction affects each magnetic hyperfine structure level in a different way and so it is measurable. The intrinsic quadrupole moment, Q of Eq. (4.43), is not the relevant quantity; its component in the direction of the nuclear spin axis Q_I is. For large deformations, which are the ones of interest so far as isotope shifts are concerned, the relation between Q and Q_I (often called the spectroscopic quadrupole moment) is

$$Q_I = Q\frac{I(2I-1)}{(I+1)(2I+3)} \tag{7.8}$$

See for example, Kopfermann and Schneider (1958) or Bohr and Mottelson (1975) which deals authoritatively with the whole subject of nuclear deformation. Deformed nuclei can be thought of as having an axis of symmetry, which does not, in general, lie along the nuclear spin axis. When $I = 0$ or $\frac{1}{2}$, the precession of the axis of symmetry about the spin axis smears out the nuclear charge distribution into one with spherical symmetry, and hence $Q_I = 0$.

7.4.3. ELECTRON SCATTERING

Elastic electron scattering has been mentioned in Section 7.1. All that need be added here is that if the scattering nucleus has $I \geq 1$, then the quadrupole moment Q_I of Eq. (7.8) will be involved in the scattering of electrons just as it is in the interaction of the nucleus with orbital electrons. The effect can be enhanced by observing the scattering by polarized nuclei. More information about nuclear deformations is also obtained from inelastic electron scattering experiments.

7.4.4. HYPERFINE STRUCTURE IN MUONIC SPECTRA

Just as in optical spectra of electronic atoms, the study of hyperfine structure can give information about nuclear deformation. The muon–nuclear interaction is so strong that for very deformed nuclei second-order effects are important and so nuclear deformations can be determined even when $I = 0$ or $\frac{1}{2}$. An early example of such a study is that of Hitlin *et al.* (1970), and a more recent one is that of Close *et al.* (1978).

7.4.5. EXPERIMENTAL VALUES OF β_2

As can be seen from Eq. (4.42), the deformation contribution to $\delta\langle r^2 \rangle$ is roughly proportional to $\langle r^2 \rangle$ and the change in β_2^2 between isotopes. The contribution is small for light elements because the $\langle r^2 \rangle$ factor is small. β_2 is very small for elements with a magic number of neutrons or protons and is largest where both the number of protons and the number of neutrons lie well away from magic numbers. Changes in β_2 between isotopes are thus large and positive in the $A = 140$ to 160 region, where the neutron number is just above the magic number 82. They are also large and positive in the $A > 210$ region where the neutron number is just above the magic number 126. They are large and negative in the $A = 180$ to 210 region where the neutron number is just less than 126.

Values of β_2 for the $A = 140$ to 210 region can be found in Jänecke (1981) and values for $A > 210$ are given by Milner *et al.* (1977) and Close *et al.* (1978), the former from Coulomb excitation experiments and the latter from experiments on muonic atoms.

EXPERIMENTAL TECHNIQUES

8.1. OPTICAL ISOTOPE SHIFTS

Optical isotope shifts between isotopes that differ by two in their mass number are usually comparable with or smaller than the Doppler width of spectral lines from a low-temperature light source. This is shown in Fig. 8.1 where the curves are both for spectral lines of wavelength 500 nm. The curve labeled "width" is the Doppler width for a source temperature of 500 K; that labeled "NMS" is the normal mass shift. The labeled points are examples (where the wavelength is only roughly of the order of 500 nm in some cases) of large specific mass shifts (given by numbers) and field shifts involving one outer s electron (given by letters). Since the great majority of isotope shifts are smaller than these examples, it can be seen that isotope shifts are rarely larger than the Doppler width even for a light source run at such a low temperature as 500 K.

8.1.1. DOPPLER-LIMITED TECHNIQUES

Let us consider first of all the techniques used where the Doppler width is not avoided by the use of an atomic beam or a laser (or both). Fig. 8.1 shows that the light source should have as low a temperature as possible, which is compatible with producing enough excitation to observe the spectral line the isotope shift of which is required. The best light source for isotope shift work has been found to be the hollow cathode lamp in which the cathode is in good thermal contact with a coolant such as liquid nitrogen. This type of lamp was introduced by Schüler (1926) and can give spark as well as arc spectra with reasonably narrow lines. The electrical discharge is usually run continuously in the lamp, but it can be pulsed in order to reach higher stages of ionization. Bishop and King (1971) studied Sn IV and Edwin and King (1969) studied Ce IV with such a light source. The discharge temperature of such lamps is about 400 K and so the Doppler width almost always makes it necessary to use separated isotopes. These are costly, but a useful amount of light can be obtained from these lamps with only 1 mg or so of the element under study in the hollow cathode. If the resonance lines of a simple spectrum are to be observed, then microgram quantities may be adequate. In such cases radioactive isotopes can be studied as for instance in the work of Heilig (1961) who studied the Sr II resonance lines of various isotopes including radioactive ^{90}Sr, which has a half-life of 28 years.

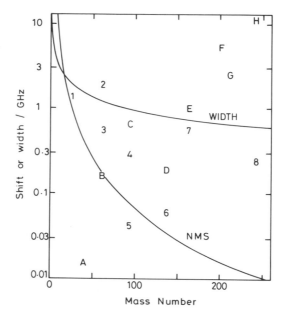

FIGURE 8.1. The variation of the Doppler width and the normal mass shift of a spectral line of wavelength 500 nm with mass number; the Doppler width is for a light source at a temperature of 500 K. The labeled points are examples of some of the larger specific mass shifts (given by numbers) and field shifts involving one outer s electron (given by letters), and are as follows:

1. *SMS* ^{26,24}Mg I, $3s^2\,^1S - 3s3p\,^3P$;
2. *SMS* ^{62,60}Ni I, $3d^84s^2\,^3F - 3d^94s\,^3D$;
3. *SMS* ^{62,60}Ni I, $3d^84s^2\,^3F - 3d^84s4p\,^5D$;
4. *SMS* ^{94,92}Mo I, $4d^55s\,^7S - 4d^55s5p\,^7P$;
5. *SMS* ^{94,92}Mo I, $4d^55s\,^7S - 4d^55p\,^7P$;
6. *SMS* ^{138,136}Ba I, $6s^2\,^1S_0 - 5d6d\,^3D_2$;
7. *SMS* ^{164,162}Dy I, $4f^{10}6s^2 - 4f^96s^25d$;
8. *SMS* ^{242,240}Pu I, $5f^6 - 5f^5$;

A. *FS* ^{41,39}K I, $4s\,^2S_{1/2} - 4p\,^2P_{1/2}$;
B. *FS* ^{62,60}Ni I, $3d^84s^2\,^3F - 3d^84s4p\,^3G$;
C. *FS* ^{94,92}Mo I, $4d^55s\,^7S - 4d^55p\,^7P$;
D. *FS* ^{138,136}Ba II, $6s\,^2S_{1/2} - 6p\,^2P_{1/2}$;
E. *FS* ^{164,162}Dy I, $4f^{10}6s^2 - 4f^{10}6s6p$;
F. *FS* ^{202,200}Hg I, $6s^2\,^1S_0 - 6s6p\,^3P_1$;
G. *FS* ^{213,211}Fr I, $7s\,^2S_{1/2} - 7p\,^2P_{3/2}$;
H. *FS* ^{243,241}Am I, $5f^77s^2 - 5f^77s7p$.

Most isotope shift measurements have concentrated on very accurate measurement of the shifts in a small number of spectral lines studied one at a time. The unwanted lines can be removed by a monochromator and the line of interest has usually been analyzed with a Fabry–Pérot etalon. The most accurate results have been obtained with photoelectric detection at the central region of the fringe pattern of a scanned etalon. An example of measurements of this type is the work of Silver and Stacey (1973a) who measured isotope shifts in Sn I. The spectral profile of one highly enriched even isotope had a total width of about 1.5 GHz and shifts between isotopes were measured with a standard deviation of about 3 MHz. Corrections had to be made for the pulls of impurity isotopes and other components, and these pulls were often about 3 MHz, but varied from 0–10 MHz. The size of the pulls was found by a computer analysis in which a theoretical profile was synthesized to match the experimental data as closely as possible, the best fit being obtained by a variational procedure. To measure shifts to better than one-hundredth of the linewidth requires careful attention to details. For example, the illumination of the etalon by different sources containing different isotopes must be kept as constant as possible. It was also found that the match between a theoretical profile and experimental data was improved when allowance was made for the change in the Doppler width between the heaviest, ^{124}Sn, and the lightest, ^{116}Sn, isotopes which were investigated.

Such attention to detail is not practical if the isotope shifts of a large number of lines in a spectrum are required. Such work, often done as an aid to the analysis of a spectrum, is often carried out with a diffraction grating, rather than an etalon, as the element of high resolution. To take the case of uranium as an example (Blaise and Radziemski, 1976), the ^{238,235}U isotope shifts range from zero up to about 20 GHz, and measurement with an accuracy of about 1 GHz is quite adequate for the identification of configurations. Much higher accuracy than this can be achieved with a grating, as shown, for example, by the isotope shifts measured in Pb I by Moscatelli et al. (1982). They used a 9.1-m focal length spectrometer of the Czerny–Turner type to measure isotope shifts in λ283.3 nm with a standard deviation of about 6 MHz. It is at shorter wavelengths such as this that the grating begins to score over the etalon because of the loss of signal that results from the multiple reflections in the latter.

Another technique which enables isotope shifts to be measured in many lines of a spectrum is Fourier transform spectroscopy. The availability of vast computing power at a reasonable cost has recently made this technique a practical possibility, and the technique has been reviewed by Brault (1982). The data used by Blaise and Radziemski (1976) on uranium was obtained partly by Fourier transform spectroscopy. Other examples of the measurement of isotope shifts by this technique are the work of Engleman and Palmer (1980) on uranium and Gerstenkorn et al. (1977) on mercury. The latter were looking for, and interpreting, the odd–even effects in the spectrum of mercury which are described and discussed in Section 9.80. The radiation from various electrodeless discharge tubes containing different mixtures of enriched mercury isotopes was studied in the 1–4 μm region using a Fourier spectrometer at Orsay. From measurements on over 100 spectral lines, the isotope shifts in about 30 levels were found with a precision of about 20 MHz.

Engleman and Palmer (1980) used the Fourier Transform Spectrometer at Kitt Peak to study the uranium spectrum produced in an electrodeless discharge tube filled with ^{238}U and ^{234}U. The use of a mixed-isotope lamp prevented the measurement of small isotope shifts, but the known intensity ratio of the two isotopes was used as an aid in the identification of uranium spectral lines. Isotope shifts were measured in over 2000 lines of U I in the visible and near-ir regions of the spectrum. Compared with the work of Gerstenkorn et al. (1977), the emphasis was on quantity rather than quality; the shifts were measured with a precision of about 100 MHz in this case.

Higher precision could have been obtained from a hollow cathode discharge, but insufficient ^{234}U was available for this type of light source to be used.

The above examples show what can be done with Doppler-limited techniques if enough care is taken; isotope shifts in the visible part of the spectrum can be measured to about 100 MHz in thousands of lines, to about 20 MHz in hundreds of lines, and to about 3 MHz in a few lines. Many of the transitions with interesting isotope shifts lie in the UV region of the spectrum. The precision with which these shifts can be measured tends to get less as the wavelength shortens, mainly because the Doppler width is proportional to the frequency of the transition. The Doppler width $\Delta \nu_D$, as a fraction of the transition frequency ν, is given by

$$\frac{\Delta \nu_D}{\nu} = 7.16 \times 10^{-7} \left(\frac{T}{A} \right)^{1/2} \tag{8.1}$$

where T is the temperature of the source (K) and A is the mass number of the element being studied. Experimental techniques of high-resolution classical optical spectroscopy, many of them suitable for the measurement of isotope shifts, are described in a chapter by Heilig and Steudel (1978) in the two volumes on atomic spectroscopy by Hanle and Kleinpoppen.

8.1.2. DOPPLER-FREE TECHNIQUES

The Doppler width can be considerably reduced by using an atomic beam and studying its absorption of light from a suitable source, instead of studying emission spectra as described in the previous section. The "suitable source" used nowadays is invariably a tunable dye laser but some isotope shifts were measured by absorption in atomic beams without the use of a laser. One of the earliest examples was the work of Jackson and Kuhn (1938) on the potassium resonance lines. They used an electrodeless discharge tube containing potassium and helium to give the background light and the collimation of the atomic beam was adjusted to reduce the Doppler width by a factor of about 30. This made it difficult to obtain sufficient intensity of absorption by the rare isotope ^{41}K, but even so they were able to measure the isotope shifts with an uncertainty of about 10 MHz in the two resonance lines with the aid of a Fabry–Pérot etalon. This etalon was external to the atomic beam, but Jackson (1961) used a spherical Fabry–Pérot etalon with the atomic beam passing through the etalon. The background light, in this case from a hollow cathode light source, interacted many times with the atomic beam as it underwent multiple reflections at the etalon plates, and so the amount of absorption was increased by a factor roughly proportional to the finesse of the reflective coatings on the etalon plates. Brandt et al. (1978) used the ability of this technique to give measurable absorption from very small amounts of beam source material to measure isotope shifts in the Ca I resonance line using beams of enriched calcium isotopes. They measured isotope shifts with an uncertainty of about 3 MHz. As they needed less than 1 mg of enriched calcium isotope per hour, they were able to study six calcium isotopes, not just the abundant ones.

Brandt *et al.* (1978) used a hollow cathode light source, but nearly all the recent measurements of isotope shifts by absorption in an atomic beam have used a tunable dye laser as the background light. The main advantage of the laser source is that it is so intense that even very rare isotopes can be detected without the need for enriched isotopes. The absorption is not observed directly in the laser beam. The usual method of detection is by the subsequent resonance fluorescence, the technique being known as laser-induced atomic beam fluorescence spectroscopy. The measurement of isotope shifts in Ba I is a good example of this technique because experiments were done by two independent groups (Baird *et al.*, 1979; Bekk *et al.*, 1979) on the same transition, $\lambda 553.5$ nm $6s^2\,{}^1S_0 - 6s6p\,{}^1P_1$. The discrepancies between the two sets of isotope shifts vary from 0.2 MHz to 2.2 MHz suggesting that this technique can measure isotope shifts with an uncertainty of about 1 MHz. It is much easier to measure the change in frequency or wavelength of the tunable laser than to try to measure the shift between the resonance fluorescence of the different isotopes. There are various ways of doing this: the Oxford group (Baird *et al.*, 1979) sent some of the light from the dye laser through a reference etalon which was pressure-scanned and locked to the dye laser. A fundamental reference frequency from a helium–neon laser, stabilized by the saturated absorption in iodine vapor, was also shone into the reference etalon and gave frequency markers about 39 MHz apart. The Karlsruhe group (Bekk *et al.*, 1979) used two atomic beams, each of a different isotope. The resonance fluorescence of each was detected with the aid of a tunable dye laser. The difference in frequency between the two lasers was measured by a heterodyne technique which had a tuning range from -220 MHz to $+5$ GHz relative to the ^{138}Ba atomic beam which was used as the reference beam. As the resonance fluorescence could be detected with less than 1 ng of barium in the oven producing the atomic beam, it was possible to measure the shift between ^{138}Ba and several radioactive isotopes.

This optical heterodyne technique has been reviewed by Griffith (1982). It was first applied to the measurement of isotope shifts by Shafer (1971) who obtained beat frequencies between stabilized xenon gas lasers filled with different isotopes of xenon. Isotope shifts in samarium have been measured by optical heterodyne spectroscopy by Griffith *et al.* (1977; 1979) who used tunable dye lasers. Unlike Bekk *et al.* (1979) they used a single atomic beam and studied all the isotopes present in a natural mixture of the samarium isotopes. They were able to measure isotope shifts with an uncertainty of 200 KHz, including ^{144}Sm, of which the natural mixture of isotopes contains only 3%. The method of detection, as in the laser experiments already described, was to look at the resonance fluorescence perpendicular to the mutually perpendicular laser and atomic beams. This method limits the transitions in which isotope shifts can be measured to those for which the lower level is populated in the atomic beam, namely the ground level or thermally populated excited levels. Sm I has seven low-lying levels, all belonging to the ground term $4f^7 6s^{2\,7}F$, and shifts were measured in transitions involving all of them (Griffith *et al.*, 1981; New *et al.*, 1981).

It is sometimes of interest to measure isotope shifts in a transition of which the lower level is not sufficiently low-lying to be populated in an ordinary atomic beam. Champeau and Keller (1978) excited atoms of krypton by passing them through an electrodeless discharge. The $4p^5 5s$ metastable levels, which lie about 80,000 cm^{-1} above the $4p^6\,^1S_0$ ground level, were produced in a beam in sufficient numbers for them to be excited by a beam of laser light and their resonance fluorescence to be detected in a similar way to that described already. Isotope shifts were measured in three transitions with an uncertainty of about 3 MHz. Brand *et al.* (1981) excited atoms of europium with an arc discharge between the oven and the atomic beam region. $4f^7 5d 6s$ metastable levels, which lie about 15,000 cm^{-1} above the $4f^7 6s^2$ ground level, were produced in sufficient numbers to study lines with these metastable levels as their lower levels. Isotope shifts were measured with an uncertainty of about 3 MHz. Alternatively, the atoms can be excited when in the atomic beam by one laser and a second laser used to produce further excitation and resonance fluorescence. This technique was used, for example, by Grafström *et al.* (1982) to measure isotope shifts in transitions between highly excited levels of Ba I.

The Doppler width was reduced, in the experiments described above, by crossing a laser beam with a collimated atomic beam in which the atoms travel at thermal velocities. Kaufman (1976) pointed out the advantage of having the laser and atomic beams collinear and using a fast atomic beam. The latter is produced by neutralizing an ion beam which has been accelerated through an electric potential difference. The Doppler width of the initial thermal beam remains the same in energy terms after acceleration, but it is very much reduced in velocity terms. It is the latter which is the relevant quantity so far as spectral resolution is concerned, and so isotope shifts can be measured accurately in ions and highly excited atoms. Examples of this type of experiment are the measurement of isotope shifts in Ba II by Van Hove *et al.* (1982) with an uncertainty of about 300 KHz, in Rb I by Klempt *et al.* (1979) with an uncertainty of about 8 MHz between radioactive isotopes, and in Ar II, Cl II, and S II by Eichhorn *et al.* (1982) with an uncertainty of about 15 MHz.

The techniques described so far cannot be used at wavelengths shorter than those which can be produced conveniently at an adequate intensity from a dye laser. The isotope shift measurements of Zaal *et al.* (1980) are close to where this limit lies at present. They studied numerous lines of Dy I with wavelengths down to 439 nm, at which wavelength they measured isotope shifts with an uncertainty of 2 MHz. Shorter wavelengths than this can be studied with the aid of frequency-doubling. Isotope shifts have been measured by this method by various people, often in order to study high-lying Rydberg levels. An early example was the work of Barbier and Champeau (1980) who not only used a frequency-doubled laser beam, but also used a discharge to populate a metastable level of ytterbium at the exit of their atomic beam oven. With this combination they were able to measure isotope shifts with an uncertainty of about 30 MHz in transitions involving principal quantum numbers up to 53. These lie over 30,000 cm^{-1} above the metastable level $4f^{14}6s6p\,^3P_0$, which itself lies 17,000 cm^{-1} above the ground level $4f^{14}6s^2\,^1S_0$. They are close to the ionization limit and detection was not of resonance fluorescence but of ions resulting from field ionization, a technique described by Ducas *et al.* (1975). More recent examples of the measurement of isotope shifts with a frequency-doubled laser beam are the study of $\lambda293.3$ nm of In I by Eliel *et al.* (1981), of $\lambda267.6$ nm of Au I by Kluge *et al.* (1983), and of $\lambda283.3$ nm of Pb I by Thompson *et al.* (1983). They measured isotope shifts with uncertainties of about 1 MHz, 200 MHz, and 5 MHz, respectively. The work on gold involved the radioactive ^{195}Au. This has the fairly long half-life of 183 days, and so the isotope shifts could be measured off-line, i.e., away from the place of manufacture of the radioactive isotope.

In contrast to this, in order to measure isotope shifts of short-lived isotopes, it is necessary to work on-line; short-lived in this context meaning relative to the time required to do an experiment, not relative to the lifetime of the upper atomic level of the transition being studied. The subject has been reviewed by Kluge (1978) in his article on optical spectroscopy of short-lived isotopes, by Jacquinot and Klapisch (1979), and by Otten (1981). The small number of atoms available makes it almost essential to use laser excitation to get a detectable signal.

Isotope (and often isomer) shifts have been measured on-line for the following elements (further details of the results obtained are given in the relevant sections of Chapter 9):

Sodium, ^{20}Na to ^{31}Na, $\lambda590$ nm: Huber *et al.* (1978); Touchard *et al.* (1982b).

Potassium, ^{38}K to ^{47}K, $\lambda770$ nm: Touchard *et al.* (1982a).

Rubidium, ^{76}Rb to ^{98}Rb, $\lambda420$ nm: Klempt *et al.* (1979); $\lambda780$ nm: Thibault *et al.* (1981a).

Caesium, ^{118}Cs to ^{145}Cs, $\lambda852$ nm: Thibault *et al.* (1981b).

Gold, work in progress: Kluge *et al.* (1983).

Mercury, ^{181}Hg to ^{191}Hg, $\lambda254$ nm: Bonn *et al.* (1976); Kühl *et al.* (1977); Dabkiewicz *et al.* (1979).

Francium, ^{208}Fr to ^{213}Fr, $\lambda720$ nm: Liberman *et al.* (1980).

Americium, ^{241,240m}Am, $\lambda641$ nm: Bemis *et al.* (1979).

Even with laser excitation, the method of detection has to be very sensitive in these experiments. More or less the same method was used for all the measurements on the alkalis and it was described in detail by Huber *et al.* (1978). The method detects optical resonances by a nonoptical technique employing magnetic deflection. If the laser is tuned to excite ground level atoms from one hyperfine sublevel then, in general, they can return by fluorescence to either the same or another hyperfine sublevel of the ground level. The laser beam is strong enough for this process to pump the atoms from one hyperfine sublevel to another within the ground level. If this is done in a magnetic field the effect of the pumping is to change the orientation of the nuclei. The experiment is arranged so that the atomic beam passes rapidly into a strong field region where I and J are decoupled. It is also arranged that the strong field focuses atoms with $m_J = +\frac{1}{2}$ and defocuses those with $m_J = -\frac{1}{2}$. Resonance is detected as a change in the number of focused atoms after they have been ionized and passed through a mass spectrometer to give the required isotope only.

A nonoptical method of detection called RADOP was used for the odd mercury isotopes (Huber *et al.*, 1976). This also involved hyperfine optical pumping as described above, but the orientation of the nuclei was monitored by the asymmetry of their β radiation; RADOP standing for radiation detected optical pumping. The pumping was done by light from a mercury lamp. The even mercury isotopes have no nuclear spin, being even–even nuclides, and show no asymmetry; in their case laser excitation was used, 254 nm being reached by frequency-doubling, and resonance detected by observing the resonance fluorescence (Duke *et al.*, 1977).

So far in this section the Doppler width has been avoided, or at least reduced, by using an atomic beam; the advent of the laser has opened up other ways of doing Doppler-free spectroscopy which rely on multiple-beam laser techniques. These techniques, which have been reviewed by Demtröder (1978) and Cagnac (1982), have been used to measure isotope shifts in some cases. The two most popular techniques are saturation spectroscopy and Doppler-free two-photon spectroscopy.

In saturation spectroscopy a laser beam is split and sent through an absorbing cell in opposite directions. The intense monochromatic saturating beam depopulates the lower level of some of the absorbing atoms so that the probe beam, passing through in the opposite direction, is not absorbed. This happens for only those atoms that have no velocity in the direction of the laser beams and so there is no Doppler broadening. Brechignac (1977) was one of the first to measure isotope shifts with this technique. Shifts in $\lambda 556$ nm and $\lambda 557$ nm of Kr I were measured with an uncertainty of about 1 MHz. A more recent example is the measurement by Coillaud *et al.* (1980) and Wagstaff and Dunn (1980) of isotope shifts in $\lambda 297$ nm of Hg I with uncertainties of 5 or 10 MHz. The lower level of the transition, $6s6p\,^3P_0$, was

TABLE 8.1

Some Examples of Isotope Shifts That Have Been Measured by Doppler-Free Two-Photon Spectroscopy

Isotopes	Two-photon transition	Laser wavelength (nm)	Isotope shift (MHz)	Reference
^{2,1}H	$1s\,^2S_{1/2}-2s\,^2S_{1/2}$	486	670992(6)	Wieman and Hänsch (1980)
^{4,3}He	Various (e.g., $2s\,^3S_1-4s\,^3S_1$)	657	42906(4)	de Clercq et al. (1981)
^{22,20}Ne	Various (e.g., $3s[\tfrac{3}{2}]_2-4d'[\tfrac{5}{2}]_2$)	592	2782(2)	Giacobino et al. (1979)
^{40,39}K	Various (e.g., $4s\,^2S_{1/2}-13s\,^2S_{1/2}$)	587	330(4)	Pendrill and Niemax (1982)
^{48}Ca to ^{40}Ca (e.g., ^{44,42}Ca)	$4s^2\,^1S_0-4s5s\,^1S_0$	600	563.7(2)	Palmer et al. (1982)
^{138}Ba to ^{134}Ba (e.g., ^{138,136}Ba)	Various (e.g., $6s^2\,^1S_0-6s9s\,^1S_0$)	537	36(1)	Jitschin and Meisel (1980)

populated by a mild discharge in the absorption cell, and the UV wavelength of λ297 nm was produced by frequency-doubling.

In Doppler-free two-photon spectroscopy atomic transitions are induced by the simultaneous absorption of two photons by one atom. The intensity of the laser field has to be high and the Doppler width is eliminated by arranging for the two photons to be travelling in opposite directions by directing the output from the laser into two beams travelling in opposite directions. The Doppler shifts of the two beams are equal and opposite, and so all atoms, irrespective of their velocity, absorb an energy $2\,h\nu$ in a two-photon absorption where ν is the frequency of the laser light. There is a Doppler-broadened background from absorption of two photons travelling in the same direction but techniques are available to reduce this. The absorption is usually detected by observing fluorescence. Some examples of isotope shifts measured by this technique are given in Table 8.1. The two most recent papers are notable for the great accuracy obtained in the case of calcium, and for the use of a radioactive isotope ^{40}K, in the case of potassium. All of the work mentioned in Table 8.1 is notable for being mentioned in Chapter 3 or Chapter 9; none of the isotope shifts were measured just to demonstrate a new spectroscopic technique.

There are other techniques of laser spectroscopy [see the review by Ferguson (1982)] that can be used to measure isotope shifts. Two pieces of work that are mentioned in Chapter 9 are the work of Freeman et al. (1980) on helium and Siegel et al. (1981) on molybdenum. They used the technique of intermodulated fluorescence and optogalvanic spectroscopy to measure isotope shifts in helium with an uncertainty of 15 MHz and in molybdenum with an uncertainty of 2 MHz. The latter experiment even used a hollow cathode, so it would seem that a device first used for the measurement of isotope shifts in the 1920s still has some life left in it, even if in a somewhat modified form.

The present situation is that Doppler-free techniques allow isotope shifts to be measured with an uncertainty of a few MHz in a large number of transitions, not just resonance lines, between many isotopes including those of low abundance and even radioactive ones. In a few cases it has been found possible to reduce the uncertainty to a few hundred KHz.

8.2. ELECTRONIC X-RAY ISOTOPE SHIFTS

The majority of the isotope shift measurements in x-ray spectra of electronic atoms have been carried out at the California Institute of Technology; their experimental methods have been described by Chesler and Boehm (1968) and Lee and Boehm (1973). Most of the remaining measurements have been carried out at the Delft University of Technology; their experimental methods have been described by van Eijk and Schutte (1970). The precision obtained by the two groups is very similar, as shown in the case of the field shift for ^{164,162}Dy in the $K\alpha_1$ transition, which has been measured by both. Bhattacherjee *et al.* (1969) obtained 55.7(4.3) meV while van Eijk and Schutte (1970) obtained 58.1(3.3) meV. In a few cases the uncertainty in the field shifts has been reduced to less than 2 meV: ^{124,116}Sn, 35.0(1.3) meV (Chesler and Boehm 1968); ^{142,140}Ce, 51.5(1.9) meV (van Eijk and Visscher 1970); ^{116,110}Cd, 28.7(1.3) meV (van Eijk *et al.*, 1979). It should be stressed that the above uncertainties are experimental in origin. The uncertainty in the field shift is the same as the uncertainty in the measured isotope shift for an x-ray transition. This is unlike the case of an optical transition where the uncertainty in our knowledge of the mass shift can be far larger than the uncertainty in the measured isotope shift.

8.3. MUONIC X-RAY ISOTOPE SHIFTS

As in the case of x-ray isotope shifts in electronic atoms, the shifts in muonic atoms have been measured by only a few groups who have access to the necessary facilities, and so the experimental details of the techniques used will not be described here. As explained in Section 7.2, a measurement of muonic atom transition energies for one isotope can give information about the nuclear charge distribution of that isotope in a way that is impossible for electronic atoms. It is thus of interest to measure absolute transition energies in muonic atoms as well as the differences between isotopes. The latter are useful, however, because to extract nuclear sizes from the energy measurements it is necessary to make various corrections, and the uncertainty in our knowledge of these corrections is very much reduced if differences between isotopes rather than absolute values are considered. Also, some experimental uncertainties, such as absolute energy calibration of x-rays in the MeV region, do not show up in the measurement of isotope shifts. The uncertainty in the nuclear parameters obtained from muonic x-ray energies arises mainly from the experimental uncertainties in the light elements, and mainly from the uncertainties in the corrections in the case of the heavy elements. The transition of most use, so far as comparison with isotope shifts in electronic atoms is concerned, is the $1s_{1/2}-2p_{3/2}$ transition. Table 8.2 gives some recently measured values of muonic atom energies for this transition, both for individual isotopes and also for the shift between pairs of isotopes. It can be seen that a large range of transition energies is involved, going from less than 100 keV to over 5000 keV. For the heaviest elements there is no point in measuring an isotope shift rather than the energies of separate isotopes because any gain in experimental precision would be masked by the uncertainty in the nuclear polarization correction. For example, in the case of

TABLE 8.2

$1s_{1/2}-2p_{3/2}$ Muonic X-Ray Energies and Isotope Shifts: Some Recently Measured Values for Elements Spanning the Periodic Table

Isotope	Transition energy (keV)	Isotope pair	Isotope shift (keV)	Reference
12	75.262(5)	^{14,12}C	0.093(3)	Schaller et al. (1982)
^{44}Ca	783.156(26)	^{44,42}Ca	−0.213(16)	Wohlfahrt et al. (1981)
^{52}Cr	1092.286(21)	^{54,52}Cr	−2.398(20)	Wohlfahrt et al. (1981)
^{96}Mo	2714.635(70)	^{98,96}Mo	−6.189(35)	Schellenberg et al. (1980)
^{134}Ba	3992.66(16)			Shera et al. (1982)
^{136}Ba	3991.01(15)			Shera et al. (1982)
^{149}Sm	4491.63(60)[a]	^{152,149}Sm	38.18(55)	Barreau et al. (1981)
^{188}Os	5540.30(55)[a]			Hoehn et al. (1981)
^{190}Os	5533.30(53)[a]			Hoehn et al. (1981)
^{202}Hg	5824.08(52)			Hahn et al. (1979)
^{204}Hg	5815.83(52)			Hahn et al. (1979)

[a] Energy corrected for nuclear polarization, but the uncertainty in this correction is not included.

mercury, the nuclear polarizations are about 7 keV and differ from isotope to isotope. Ignoring the uncertainty in the nuclear polarization, the experimental uncertainties in the measured isotope shifts of Table 8.2 range from less than one, to over ten parts in a hundred. This is very similar to the comparable figure for optical isotope shifts where a typical shift is about 500 MHz, and, as described in Section 8.1.2, these can often be measured with an uncertainty of a few MHz or less. A significant difference between the two techniques is in their scope. The measurement of muonic x-ray energies requires a few grams of a highly enriched isotope, and so measurements can only be made on the stable and abundant isotopes. Optical isotope shifts, on the other hand, can be measured with natural mixtures of isotopes, or with less than, sometimes much less than, a milligram of an enriched or radioactive isotope.

ISOTOPE SHIFTS AND OTHER RELEVANT WORK
AN ELEMENT BY ELEMENT REVIEW

No new experimental results are presented in this chapter, but in some cases the analysis of the results is new. Not every recent paper is mentioned, but I have tried to include all that seemed to me to be of major importance.

9.1. HYDROGEN

See Section 3.1; the isotope shift is entirely the normal mass shift. Extremely accurate results are given in a recent paper by Amin *et al.* (1981).

9.2. HELIUM

Comparison with theory reveals the presence of small relativistic effects in the mass shifts (see Section 3.2.1). Reference to Table 3.2 shows that the largest specific mass shift occurs in the ground term $1s^2\,^1S$. Transitions to this term involve the vacuum-UV region of the spectrum and no interferometric or laser measurements have yet been made. Herzberg (1958) measured a shift in the 58.43 nm line of 264(1) GHz between ^{4}He and ^{3}He. The normal mass shift is 230 GHz, so, ignoring the minute field shift, the specific mass shift is 34 GHz. This is in good agreement with the calculated value of 33.35 GHz. Herzberg used a 3-m vacuum grating spectrograph in about tenth order, and improved the precision with which the ionization potential of helium was known by a factor of 100.

9.3. LITHIUM

9.3.1. Li II

The earliest work on the isotope shift in a two-electron spectrum was done on Li II (the spectrum of Li^+) as described in Section 3.2. Recent work, in which a field shift was detected alongside the much larger mass shift, is discussed in Section 3.2.2. This work shows the need for care when measuring an isotope shift when magnetic nuclear hyperfine structure (hfs) is present.

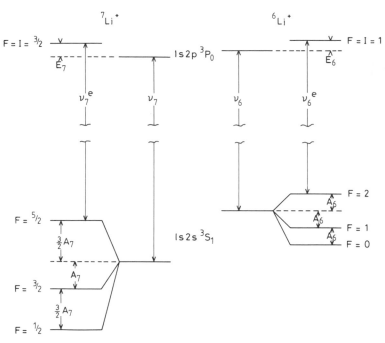

FIGURE 9.1. Hyperfine structure and isotope shift in Li$^+$. The experimentally measured isotope shift $(\nu_7^c - \nu_6^c)$ is 26,618(2) MHz. The magnetic hyperfine structure constants are $A_6 = 3001.8$ MHz and $\frac{3}{2}A_7 = 11,890.4$ MHz. The perturbations of $2\,^3P_0$ by the hyperfine structure of the $2\,^3P_1$ and $2\,^3P_2$ levels are $E_6 = 63$ MHz and $E_7 = 804$ MHz. The isotope shift allowing for perturbation $(\nu_7 - \nu_6)$ is 34,766(3) MHz. The figure is not to scale and the data are from Bayer et al. (1979), updated by a private communication.

Both the stable isotopes of lithium have a nuclear spin; for ^{7}Li $I = \frac{3}{2}$ and for ^{6}Li $I = 1$. The transitions $1s2s\,^3S_1$–$1s2p\,^3P$ were studied, and, to avoid hfs in the upper level, the isotope shift was measured in the $2\,^3S_1$–$2\,^3P_0$ transition. The lower level has a triplet hfs and the frequency difference between $2\,^3S_1$ $(F = \frac{5}{2}) - 2\,^3P_0$ of ^{7}Li and $2\,^3S_1$ $(F = 2) - 2\,^3P_0$ of ^{6}Li was measured and found to be 26,618(2) MHz. The isotope shift between the centers of gravity of $2\,^3S_1$ and $2\,^3P_0$ between ^{7}Li and ^{6}Li is obtained from a knowledge of the magnetic hfs constants of $2\,^3S_1$ for ^{7}Li and ^{6}Li, this shift has the value 35,507 MHz as can be seen from Fig. 9.1. This, however, is not the true isotope shift, since, although the level $2\,^3P_0$ has no hyperfine splitting, its position is disturbed because of hfs. This is because the hfs in the $2\,^3P$ term is not very small compared with the fine structure in the term; the hfs is of the order of 10 GHz and the fine structure is of the order of 80 GHz. The $2\,^3P_0$ level cannot be considered in isolation from the rest of the $2\,^3P$ term, because the hfs perturbs $2\,^3P_0$. These anomalous coupling conditions in Li$^+$ have long been known (Güttinger and Pauli, 1931). They have also been observed in other alkaline earth spectra. Bayer et al. (1979) obtained analytical relations between the hfs interaction constants, the (undisturbed) fine structure splittings, and the energy differences from the diagonalization of the energy matrix. The effects are quite different in the two isotopes (E_7 and E_6 in Fig. 9.1), so the isotope shift in $2\,^3S_1 - 2\,^3P_0$ between ^{7}Li and ^{6}Li is 34,766(3) MHz.

9.3.2. Li I

The optical isotope shifts in this spectrum are discussed in Section 3.3.1. The agreement between experiment and theory is very good, but not perfect. There is still useful work to be done on the calculation of specific mass shifts in three-electron spectra.

9.4. BERYLLIUM

Beryllium has only one stable isotope and no isotope shifts have been measured in the spectra of beryllium. Some muonic x-ray transitions of ^{9}Be have been studied by Schaller et al. (1980). Comparison with calculated energies for a point nucleus shows that about 2 parts in 1000 of the measured energy in the 1s–2p transition is due to the finite size of the nuclear charge distribution. This leads to a value of 5.71(81) fm^2 for $\langle r^2 \rangle$, the mean square radius of the nuclear charge distribution, in agreement with the more accurate result deduced from electron scattering experiments.

9.5. BORON

Olin et al. (1981) deduced from muonic x-ray transitions that the mean square radius of the nuclear charge of ^{11}B is 5.66(19) fm^2 and that of ^{10}B is 5.95(29) fm^2.

9.5.1. B II

The following ^{11,10}B isotope shift has been measured by Burke (1955): $2s2p\,^1P_1 - 2p^2\,^1D_2$; 26.0(8) GHz ($NMS = 4.3$ GHz). The field shift being negligible in comparison, the specific mass shift is 21.7(8) GHz. A value for the specific mass shift of 36 GHz was calculated by Bauche (1969) using Hartree–Fock wavefunctions.

9.5.2. B I

Burke (1955) measured the following ^{11,10}B isotope shift: $2s^2 2p\,^2P - 2s^2 3s\,^2S$; $-4.1(2)$ GHz ($NMS = 6.0$ GHz). Edlen et al. (1970) obtained, amongst other shifts: $2s2p^2\,^2D - 2s^2 4f\,^2F$; $-18.4(3)$ GHz ($NMS = 1.8$

GHz). The specific mass shift in the $^2P-^2S$ transition is $-10.1(2)$ GHz which is to be compared with Bauche's (1969) calculated value of -13 GHz. Edlen *et al.* (1970) pointed out that in the transition from ground term to ionization limit, $2s^2 2p\,^2P-B^+\,2s^2$, the normal mass shift of 10.0 GHz is cancelled out by an equal and opposite specific mass shift to give an isotope shift of approximately zero. They used the flash-photolysis technique to obtain absorption photographic plates of the $2s2p^2\,^2D-2s^2nf\,^2F$ lines. The two-electron transition arises from the substantial sharing of properties between $2s2p^2\,^2D$ and $2s^2 3d\,^2D$.

9.6. CARBON

Bauche (1969) has calculated, using Hartree–Fock wavefunctions, that the specific mass shift k factor for the transition $2p^2\,^1S-2p3s\,^1P$ is -0.43175 a.u. The very first specific mass shift calculations for any element using Hartree–Fock wavefunctions were made by Nicklas and Treanor (1958) for carbon and oxygen.

Schaller *et al.* (1982) have measured muonic x-ray energies in carbon isotopes, and deduce the following changes in the mean square radius of the nuclear charge distribution: $\delta\langle r^2\rangle = 0.04(5)$ fm^2 for $^{13,12}C$ and $0.12(5)$ fm^2 for $^{14,12}C$. Ruckstuhl *et al.* (1982) have made very high precision measurements of muonic x-ray energies in ^{12}C and deduce that $\langle r^2\rangle = 6.166(10)$ fm^2.

9.7. NITROGEN

Schaller *et al.* (1980) deduced from muonic x-ray transitions that the mean square radius of the nuclear charge distribution is $6.55(5)$ fm^2.

9.8. OXYGEN

Bauche (1969) calculated, using Hartree–Fock wavefunctions, that the k factor for the specific mass shift in the $2p^3 3s\,^5S-2p^3 3p\,^5P$ transition is 0.05934 a.u.

The isotope shift has been measured in some muonic x-ray transitions by Backenstoss *et al.* (1980). They found a field shift in the 1s level of $-80(5)$ eV between ^{18}O and ^{16}O. They used the value of the nuclear radius of ^{16}O, which is known more accurately from electron scattering results than from muonic x-ray transition energies, to find the change in radius between ^{18}O and ^{16}O. They found that the change in the mean square radius of the nuclear charge distribution, $\delta\langle r^2\rangle = 0.42(3)$ fm^2.

9.9. FLUORINE

Fluorine has only one stable isotope and no isotope shift work has been done on the spectra of fluorine. Schaller *et al.* (1978) have studied some muonic x-ray transitions of ^{19}F. Comparison with calculated energies for a point nucleus shows that about 15 parts in 1000 of the measured energy in the $1s-2p_{1/2}$ transition is due to the finite size of the nuclear charge distribution. This leads to a value of 8.40(6) fm^2 for the mean square radius of the nuclear charge distribution. This was in good agreement with the result deduced from electron scattering experiments which had a similar uncertainty.

9.10. NEON

The mass shifts in the optical spectrum of neon have been discussed in Section 3.3.2. A line that has been much studied by laser physicists is one of the lasing transitions of the He–Ne laser, λ632.8 nm $2p^5(^2P_{1/2})3p[\frac{3}{2}]_2-2p^5(^2P_{1/2})5s[\frac{1}{2}]_1$. Isotope shifts have been measured by Ballik (1972) and Kotlikov and Tokarev (1980), for example, but the interest has been more in laser processes than in the isotope shifts themselves. The shift is, in fact, 889(9) MHz for ^{22,20}Ne. Lyons *et al.* (1981) and Hansch *et al.* (1981) have used neon to show that radiofrequency detection with intermodulated, or polarization intermodulated, excitation spectroscopy provides a sensitive and convenient tool for high-resolution laser spectroscopy.

9.11. SODIUM

Sodium (having but one stable isotope) isotope shift measurements inevitably involve the use of radioactive isotopes. Huber *et al.* (1978) studied the D$_1$ line of all isotopes from ^{21}Na to ^{31}Na. Twenty-GeV protons bombarded a uranium target to produce the sodium isotopes. The target was heated so that a thermal atomic beam of sodium was produced. This was crossed by a tunable cw dye laser, and optical resonance was detected through an optical pumping process, a mass spectrometer being used to ensure isotopic selection. The normal mass shift for ^{25,23}Na is 971 MHz and an educated guess suggests that the field shift is about 6 MHz. It follows from the measured shift that the specific mass shift is about 370 MHz. Hartree–Fock calculations (Bauche, 1974) gave the significantly smaller value of 60 MHz. This and other earlier work is summarized and discussed by Touchard *et al.* (1982b) who extended the work to include ^{20}Na. Lindroth and Mårtensson-Pendrill (1983) have evaluated the specific mass shift taking into account second-order pair correlation effects as was done for the case of lithium (see Section 3.3.1). They found this brought the theoretical evaluation into much better agreement with the experimental result in the resonance line. The figure of 6 MHz for the field shift is arrived at from a knowledge of the nuclear quadrupole deformation (determined from the hfs B ($^2P_{3/2}$) factor), and the assumption that (see Chapter 10)

$$\delta\langle r^2 \rangle_{\text{volume}} = 0.29 \, A^{-1/3} \, \delta A \qquad (9.1)$$

Although the absolute value of the field shifts is not known, there is an obvious discontinuity at ^{25}Na. Between the lighter isotopes the field shift is small and positive, i.e., $\langle r^2 \rangle$ is decreasing with increasing A; between the heavier isotopes it is certainty negative, i.e., $\langle r^2 \rangle$ is definitely increasing with increasing A. The change is so large that it cannot be entirely explained by $\delta\langle r^2 \rangle_{\text{shape}}$; $\delta\langle r^2 \rangle_{\text{volume}}$ must change through the series of isotopes. The discontinuity at ^{25}Na ($N = 14$) is a shell effect occurring at the closure of the $1d_{5/2}$ subshell of neutrons.

Schaller *et al.* (1978) have measured the energies of muonic x-ray transitions in ^{23}Na, and deduce that $\langle r^2 \rangle$, the mean square radius of the nuclear charge distribution, is 8.92(5) fm^2. Electron scattering data give a similar but less precise result.

9.12. MAGNESIUM

The first spectrum of magnesium has been a testing ground for the theoretical evaluation of specific mass shifts ever since the pioneering work of Vinti (1939). Specific mass shifts have been evaluated by Bauche (1974), using Hartree–Fock wavefunctions, and by Labarthe (1973) who carried the calculation for the 3s3p configuration to second order using a multiconfigurational Hartree–Fock approach. The most recent measurements of specific mass shifts have been made by Hallstadius and Hansen, and their results as well as earlier work are discussed by Hallstadius (1979). They used a pressure scanned Fabry–Pérot interferometer to study separated isotopes excited in a hollow cathode. To within their experimental uncertainty of about 3 to 30 MHz, they found no evidence for any field shifts between the isotopes they studied: ^{26}Mg, ^{25}Mg, and ^{24}Mg. This is to be expected since electron scattering gives $\delta\langle r^2 \rangle$ for ^{26,24}Mg as only 0.1 fm^2, and so the normal mass shift can be subtracted from the measured shifts to give the specific mass shift. The agreement between theory and experiment is good and is shown in Table 9.1, the figures for which have been taken from Hallstadius (1979) and include experimental results of Kelly (1957).

TABLE 9.1
Specific Mass Shifts for ^{26,24}Mg in Mg I

Transition	Theory (MHz)	Experiment (MHz)
$3s^2\,^1S_0-3s3p\,^1P_1$	-630^a	$-435(21)$
$3s^2\,^1S_0-3s3p\,^3P_1$	$+1230^a$	$+1332(21)$
$3s3p\,^3P-3s3d\,^3D$		$-1317(3)$
$3s3p\,^3P-3s3p\,^1P$	-1860^a	$-1767(30)$
	-1893^b	

^{a}From Bauch (1974).
^{b}From Labarthe (1973).

9.13. ALUMINUM

Aluminum has only one stable isotope, and there is no work on isotope shifts to report. Schaller et al. (1978) have measured the energies of muonic x-ray transitions and deduce that the mean square radius of the nuclear charge distribution of ^{27}Al is 9.35(3) fm^2. Electron scattering data give a less precise result in agreement with this.

9.14. SILICON

Schaller et al. (1978) deduced from their measurements of muonic x-ray transition energies that the mean square radius of the nuclear charge distribution $\langle r^2 \rangle$, is 9.79(2) fm^2 for the natural mixture of isotopes (mainly ^{28}Si). Electron scattering data for individual isotopes do not show any change in $\langle r^2 \rangle$ between isotopes.

9.15. PHOSPHORUS

Phosphorus has only one stable isotope and there is no isotope shift work to report. Work on muonic x-ray transitions (Schaller et al., 1978) shows that $\langle r^2 \rangle$ is 10.16(2) fm^2. Electron scattering data give a less precise result in agreement with this.

9.16. SULPHUR

Natural sulphur is 95% ^{32}S and Schaller et al. (1978) found from measurements of muonic x-ray transition energies that $\langle r^2 \rangle$, the mean square radius of the nuclear charge distribution, of the natural mixture of isotopes is 10.65(1) fm^2.

The isotope shift for ^{34,32}S has been measured in the optical transition $3p^2(^3P)3d\,^4F_{9/2}-3p^2(^3P)4p\,^4D_{7/2}$ of S II by Eichhorn *et al.* (1982). Resonance absorption between an ion beam of sulphur and a cw dye laser was detected by observing the fluorescence. The Doppler width was reduced by the collinear laser spectroscopy technique in which the ion beam was accelerated through a potential of 100 V. The measured isotope shift was $-2178(38)$ MHz compared with the normal mass shift of 539 MHz. The field shift is negligible compared with these figures, so the specific mass shift is $-2717(38)$ MHz. The method of Vinti (1939) was used to evaluate the specific mass shift using Hartree–Fock wavefunctions. Assuming pure Russell–Saunders coupling, the theoretical evaluated value was found to be -2915 MHz, in fair agreement with the measured value. The large size of the specific mass shift is discussed, for the similar case of argon, in Section 9.18.

9.17. CHLORINE

Eichhorn *et al.* (1982) (see Section 9.16) found the specific mass shift for ^{37,35}Cl in $3p^3(^4S)3d\,^5D_4-3p^3(^4S)4p\,^5P_3$ of Cl II was $-2589(28)$ MHz on the assumption that the field shift was negligible, this was in fair agreement with their theoretical evaluated value of -2925 MHz.

Elastic electron scattering has been observed from ^{35}Cl and ^{37}Cl by Briscoe *et al.* (1980). They deduced mean square nuclear charge-radius values of 11.48(10) fm^2 for ^{35}Cl and 11.45(10) fm^2 for ^{37}Cl.

9.18. ARGON

Isotope shifts were measured in the spectrum of argon as long ago as 1937 (Kopfermann and Kruger, 1937); this paper is a good example of the sort of work on isotope shifts done in this period. This work and other later work showed that the isotope shifts were predominantly the normal mass shift; the specific mass shift and the field shift were small in comparison. The first useful isotope shift results to be obtained using a laser were in Ar I (Brochard *et al.*, 1967) and these revealed a large specific mass shift for ^{40,36}Ar of about 840 MHz in the two ir transitions $3p^5(^2P_{1/2})4p[\frac{3}{2}]_2-3p^5(^2P_{3/2})3d[\frac{1}{2}]_1$, and $3p^5(^2P_{1/2})4p[\frac{1}{2}]_1-3p^5(^2P_{3/2})3d[\frac{1}{2}]_0$. The corresponding k factor (Eq. 3.21) is 0.08 a.u.; it can be seen from the lower half of Table 3.9 that this is quite a large value for this part of the Periodic Table. However, this result pales into insignificance compared with the specific mass shifts that have been measured in Ar II by Eichhorn *et al.* (1982). Their technique of collinear laser spectroscopy with ion beams has been briefly described in Section 9.16 and

TABLE 9.2

Specific Mass Shifts for ^{40,36}Ar in Ar II

Transition	Theory (MHz)	Experiment (MHz)
$3p^4(^1D)3d\,^2G-3p^4(^1D)4p\,^2F$	-5786	$-4941--4947$
$3p^4(^3P)3d\,^4F-3p^4(^3P)4p\,^4D$	-5696	$-4942--4974$
$3p^4(^3P)3d\,^4F-3p^4(^3P)4p\,^2D$	-5726	-4888
$3p^4(^3P)3d\,^4D-3p^4(^3P)4p\,^4P$	-6445	$-5607--5776$

more fully in Chapter 8. They observed specific mass shifts for ^{40,36}Ar of about -5000 MHz, i.e., k factors of about -0.5 a.u., in 3d–4p transitions. The shifts varied very little with J within a term, suggesting that the coupling was approximately of the Russell–Saunders type. On this assumption, Eichhorn *et al.* (1982) calculated theoretical values of the specific mass shift with Vinti integrals using Hartree–Fock wavefunctions. The major contributions were from the (3d–3p) and (3d–2p) terms but the core, particularly (3p–1s) made a significant contribution in the opposite direction. Some of the measured variation between terms of the configurations was also apparent in the *ab initio* theoretical evaluations as can be seen from Table 9.2. Even the absolute agreement was fairly good.

All the above results were based on the assumption that the field shift is negligible compared with the mass shift. This is eminently reasonable because Pfeiffer and Daniel (1976) deduced from their study of muonic x-ray energies that the mean square radius of the nuclear charge distribution $\langle r^2 \rangle$ for ^{40}Ar was 11.76 fm^2, and that the changes between isotopes, $\delta\langle r^2\rangle$, were only 0.09 fm^2 for ^{38,36}Ar and 0.1 fm^2 for ^{40,38}Ar.

9.19. POTASSIUM

Precise 2p–1s muonic x-ray transition energies have been measured in ^{41}K and ^{39}K by Wohlfahrt *et al.* (1981) and from these they deduce the following mean square radii of the nuclear charge distribution: 11.94 fm^2 for ^{41}K and 11.82 fm^2 for ^{39}K. These results are model-dependent, but they deduce a model-independent value for $\delta(\langle r^2\rangle^{1/2})$ of 0.0171(39) fm, i.e., $\delta\langle r^2\rangle = 0.12(3)$ fm^2.

Potassium has a simple alkali spectrum and so the field shift can be calculated fairly easily. Touchard *et al.* (1982a) found that for ^{41,39}K in the $4s\,^2S_{1/2}-4p\,^2P_{1/2}$ D$_1$ line the field shift was $-15(4)$ MHz based on the above value for $\delta\langle r^2\rangle$. The normal mass shift is 268 MHz so, since the measured shift is 235 MHz; the specific mass shift is $-18(4)$ MHz. A theoretical evaluation of the specific mass shift including correlation effects has been made by Mårtensson and Salomonson (1982) who obtained -23 MHz. This is in very good agreement with the experimental value, bearing in mind that the theoretical evaluation involves the partial cancellation of contributions of up to a few hundred MHz. The p shells in the core lead to core–core contributions and they give core-valence contributions for the s state. The good agreement may be fortuitous since higher-order effects could be important, but the central-field model (which predicts a far larger specific mass shift) is clearly shown to be inadequate.

Touchard *et al.* (1982a) used laser spectroscopy on a thermal atomic beam containing isotopes ranging from ^{38}K to ^{47}K. The techniques are described in the very similar experiments with sodium beams (Huber *et al.*, 1978). Having determined the field shift for ^{41,39}K as described above, the field shifts between all the other isotopes can be found and hence the changes in $\langle r^2 \rangle$ between them. There are three quite distinct regions; up to $A = 41$ $\langle r^2 \rangle$ increases more rapidly with A than it does from $A = 41$ to $A = 45$, above $A = 45$ $\langle r^2 \rangle$ decreases with A. Although the three regions are quite distinct, it is not possible to give unique values for $\delta \langle r^2 \rangle$ in each region because of the relatively large uncertainty in the value of $\delta \langle r^2 \rangle$ for ^{41,39}K upon which the analysis is based. The experimental work involved the use of atomic beams; one of the pioneering uses of these to measure an isotope shift was in potassium (Jackson and Kuhn, 1938).

The tunable dye laser has allowed isotope shift measurements to be extended, not only to radioactive isotopes as described above, but to transitions involving highly excited Rydberg states. Pendrill and Niemax (1982) have used Doppler-free two-photon resonance absorption to study n^2S and n^2D levels where n ranged from 6 to 13. Extrapolation to $n = \infty$ was possible, so that the isotope shifts of levels could be quoted relative to the ionization limit, $3p^6\,^1S_0$. The interaction of a highly excited valence electron with the rest of the atom was found to be surprisingly large. For example, in the $6\,^2$D level the valence electron is several Å from the nucleus, and yet the isotope shift in the $6\,^2$D–$\infty\,^2$D "transition" differs significantly from the normal mass shift.

9.20. CALCIUM

Calcium is perhaps the favorite light element of the nuclear physicist. It has a magic number of protons and so has many stable isotopes, namely, five plus naturally occurring but radioactive ^{48}Ca. The most abundant (and lightest) isotopes, ^{40}Ca, and ^{48}Ca, are doubly magic; they also have a magic number of neutrons. Calcium has also been a favorite element of the measurers of optical isotope shifts. For many years it was the lightest element in which the field shifts were large enough, relative to the mass shifts, for estimates of nuclear sizes to be made from optical isotope shift measurements. Its relatively simple spectrum was attractive to both experimentalists and theoreticians and it was one of the elements for which Bauche (1974) evaluated the specific mass shift.

The nuclear charge distributions of the calcium nuclei have been studied in great depth by many techniques. Two recent articles that review such work are Brown *et al.* (1979) and Träger (1981). The relevant nuclear property so far as isotope shifts are concerned is the change between isotopes in the mean square radius of the nuclear charge distribution $\delta\langle r^2 \rangle$. By coincidence, $\delta\langle r^2 \rangle$ between the doubly magic isotopes ^{48}Ca and ^{40}Ca is extremely small and so all measured isotope shifts for ^{48,40}Ca are predominantly mass shifts.

The most recent measurements of muonic x-ray energies have been made by Wohlfahrt *et al.* (1981) and the most recent optical isotope shift measurements have been made: (a) in the λ657.3 nm ($4s^2\,^1S_0 - 4s4p\,^3P_1$) transition of Ca I by Bergmann *et al.* (1980) using a tunable cw dye laser crossed with an atomic beam and detecting resonance fluorescence; (b) in the λ422.7 nm ($4s^2\,^1S_0 - 4s4p\,^1P_1$) transition of Ca I by Andl *et al.* (1982) using the same technique extended to shorter wavelength by using Stilbene-3 dye pumped by UV light; (c) in the $4s^2\,^1S_0 - 4s5s\,^1S_0$ transition of Ca I by Palmer *et al.* (1982) using two-photon laser spectroscopy with a laser wavelength of 600.1 nm; and (d) in transitions to Rydberg states ($4s^2\,^1S_0 - 4sns\,^1S_0$) by Beigang and Timmermann (1982) also using two-photon laser spectroscopy.

The $\delta\langle r^2 \rangle$ values of the nuclear charge distributions are best found by plotting the optical isotope shifts against the $\delta\langle r^2 \rangle$ values obtained from muonic x-ray energies as in Fig. 6.6. The best fit line then combines the absolute accuracy of the muonic work with the greater relative precision of the optical work; it is also possible to include isotopes studied optically but not muonically. The agreement between the two types of experiment is quite satisfactory and the values of $\delta\langle r^2 \rangle$ that follow are shown in Fig. 9.2. As work is still in progress, no attempt has been made to draw Fig. 9.2 with extreme accuracy. However, it does show the feature on which all the experimental results agree, and that is that there is a marked odd–even staggering. This phenomenon was first discovered in the optical isotope shift of heavy elements many years ago, but calcium is the lightest element in which it has been seen. Although the work of Touchard *et al.* (1982a) on potassium involved an overlapping range of neutron numbers, there is no evidence of such a marked odd–even staggering in their isotope shifts. This matter is discussed in Section 10.2.

If the field shift for ^{48,40}Ca is assumed to be zero then the isotope shift can be split up into a normal mass shift and a specific mass shift. This has been done in Table 9.3 which also includes an evaluation of the specific mass shift by Bauche (1974). Perhaps the kindest thing that can be said of the latter is that the correct direction for the $^3P - ^1P$ shift was predicted.

It is known that in alkaline earth spectra there are hyperfine interactions in second order as the hyperfine and fine structure splittings are of compara-

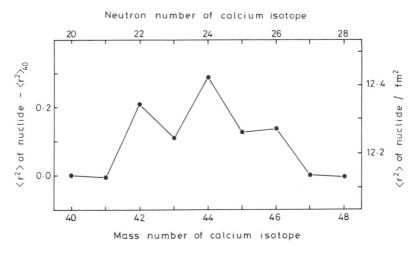

FIGURE 9.2. Mean square radii of the nuclear charge distribution of calcium isotopes. Results obtained by combining optical and muonic work drawn to show the general trends. For accurate values and uncertainties the original papers should be consulted.

ble size. This effect needs to be taken into account in precise isotope shift work as is discussed in Section 9.3.1 for the case of Li II. Palmer *et al.* (1982) point out that such corrections may need to be applied to the results discussed above in the $^1S_0 - {}^3P_1$ transition. This hyperfine-induced singlet–triplet mixing was clearly displayed in the work of Beigang and Timmermann (1982) on isotope shifts in Rydberg states involving ^{43}Ca.

9.21. SCANDIUM

Scandium has only one stable isotope, and there is no work on isotope shifts to report. Wohlfahrt *et al.* (1981) have measured the energies of muonic

TABLE 9.3
Mass Shifts for ^{48,40}Ca in Ca I

Transition	Mass shift[a] (MHz)	NMS (MHz)	SMS (MHz)	Evaluated[b] SMS (MHz)
$4s^2\,^1S_0 - 4s4p\,^3P_1$	1923	1044	-879	210
$4s^2\,^1S_0 - 4s4p\,^1P_1$	1511	1623	-112	-1260
$4s^2\,^1S_0 - 4s5s\,^1S_0$	2194	2286	-92	

[a] Assuming the field shift is zero.
[b] Theoretical evaluation by Bauche (1974).

x-ray transitions and deduce that the mean square radius of the nuclear charge distribution of ^{45}Sc is 12.60 fm^2, this being a model-dependent value.

9.22. TITANIUM

Recent work on the optical isotope shift has not yet been published. Wohlfahrt *et al.* (1981) have measured the energies of muonic x-ray transitions and deduced the following values of $\delta\langle r^2\rangle$, the change in the mean square radius of the nuclear charge distribution in fm^2: ^{50,48}Ti, $-0.164(8)$; ^{48,46}Ti, $-0.110(6)$; ^{49,48}Ti, $-0.134(25)$; ^{48,47}Ti, $-0.020(25)$. These values are model-independent and use electron scattering data for the even isotopes, hence the smaller uncertainty for the differences between them. Their model-independent value for $\langle r^2\rangle$ for ^{48}Ti was 12.95(3) fm^2.

9.23. VANADIUM

Vanadium has only one stable isotope, and there is no work on isotope shifts to report. Wohlfahrt *et al.* (1981) have measured the energy of muonic x-ray transitions and deduce a model-dependent value for the mean square radius of the nuclear charge distribution of 12.98 fm^2 for ^{51}V.

9.24. CHROMIUM

Optical isotope shifts have been measured in three $3d^5(a^6S)4s-3d^5(a^6S)4p$ transitions of Cr I (Heilig and Wendlant, 1967; Bruch *et al.*, 1969) and these can be compared with the changes in the mean square radius of the nuclear charge distribution deduced by Wohlfahrt *et al.* (1981) from muonic x-ray transitions and electron scattering experiments. The mass shift in these transitions for ^{52,50}Cr is about 80 MHz, and so since the normal mass shift is 294 MHz, the specific mass shift is about -210 MHz, i.e., the Vinti k factor is about -0.08 a.u. The model-independent values of $\delta\langle r^2\rangle/\text{fm}^2$ deduced by Wohlfahrt *et al.* (1981) were: ^{54,52}Cr, 0.342(5); ^{52,50}Cr, $-0.155(5)$; ^{53,52}Cr, 0.123(24). Their model-independent value of $\langle r^2\rangle$ for ^{52}Cr was 13.29(2) fm^2.

9.25. MANGANESE

Manganese has only one stable isotope, and there is no work on isotope shifts to report. Wohlfahrt *et al.* (1981) have measured the energy of muonic x-ray transitions and deduce a model-dependent value for the mean square radius of the nuclear charge distribution of 13.76 fm^2 for ^{55}Mn.

9.26. IRON

Optical isotope shifts have been measured (Bruch et al., 1969; Heilig and Steudel, 1974) and muonic x-ray energies have been measured (Shera et al., 1976). The following changes in the mean square radius of the nuclear charge distribution, $\delta\langle r^2 \rangle/\text{fm}^2$, were obtained: ^{58,56}Fe, 0.286(7); ^{56,54}Fe, 0.327(8); ^{57,56}Fe, 0.12. The value of $\langle r^2 \rangle$ for ^{56}Fe was 14.01 fm^2. All the above results are from the muonic work and the uncertainties (where given) were calculated to make the values model-independent, some use having also been made of electron scattering measurements (Wohlfahrt et al., 1980) in these cases.

9.27. COBALT

Cobalt has only one stable isotope, and there is no work on isotope shifts to report. Shera et al. (1976) have measured the energy of muonic x-ray transitions and deduce that the mean square radius of the nuclear charge distribution of ^{59}Co is 14.39 fm^2.

9.28. NICKEL

Some of the earliest theoretical evaluations of specific mass shifts were made by Bauche and Crubellier (1970) for atoms in the 3d series, ranging from scandium to copper. A good comparison between these evaluated shifts and measured isotope shifts can be made in the case of nickel. Steudel et al. (1980) have measured shifts in several spectral lines between five isotopes. These can be compared graphically with values of $\delta\langle r^2 \rangle$ obtained from muonic x-ray energy and electron scattering measurements (Wohlfahrt et al., 1980), using a plot similar to that of Fig. 6.6, to separate the mass shifts from the field shifts in the optical isotope shifts. The two sets of experimental results are consistent with each other, and unambiguous values for the mass shifts can be obtained. After subtraction of the normal mass shift, the experimental specific mass shifts shown in Table 9.4 were obtained. The theoretical evaluations are also given in Table 9.4, and the agreement is about what might be expected considering the relative crudity of the theoretical evaluations as compared, for instance, with what has been done in the case of potassium.

TABLE 9.4
Isotope Shifts for ^{62,60}Ni in Ni I^a

Transition	SMS (MHz)	
	Experimental value	Theoretical evaluation
$3d^94s\,^3D$–$3d^94p\,^3P$	408(27)	291
$3d^84s^2\,^3F$–$3d^84s4p\,^5D$	552(33)	270
$3d^94s\,^3D$–$3d^84s4p\,^5D$	−1446(117)	−2097
$3d^{10}\,^1S$–$3d^94p\,^1P$	−621(12)	−1104

^{a}The data are from Steudel et al. (1980).

Using the experimental values of the specific mass shifts, that is, those determined with the aid of muonic data, Steudel et al. (1980) compared their measured field shifts with results from Hartree–Fock calculations. The relative sizes of the field shifts (comparing one transition with another) showed a very satisfactory agreement between experiment and theory. This was, however, not true for the absolute values in the transition $3d^94p$–$3d^94s$; as has been found in other spectra, the ab initio Hartree–Fock values of $\Delta|\psi(0)|^2$ [see Eq. (4.24) and Eq. (4.33)] were only about two-thirds of the empirical values obtained from the use of either the Goudsmit–Fermi–Segrè formula or magnetic splittings in hfs. A more serious discrepancy, because unexplained, is that when empirical values of $\Delta|\psi(0)|^2$ are used to calculate values of $\delta\langle r^2\rangle$ from the field shifts, the values of $\delta\langle r^2\rangle$ are about 20% smaller than those found from muonic x-ray measurements. Even though the calculation involves the determination of a screening factor (see Section 4.2.2), a discrepancy of 20% in a transition as simple as $3d^94p$–$3d^94s$ is disturbingly large.

9.29. COPPER

Isotope shift work in the spectra of copper has been mainly concerned with the specific mass shift. Bauche (1966) showed that the specific mass shift was large when there was a change in the number of 3d electrons. He calculated that the Vinti k factor was about 1.5 a.u., leading to a specific mass shift for ^{65,63}Cu of about 2500 MHz, in $3d^{10}$–$3d^9$ transitions. Measurements in Cu II by Elbel et al. (1963) had shown that the specific mass shifts of the $3d^94s\,^1D$ and $3d^94s\,^3D$ levels were unequal. This is inexplicable on the simple Vinti theory (see Section 3.3) as the orbital angular momentum quantum numbers of the 3d and 4s electrons do not differ by unity. Elbel and Hühnermann (1969) explained the difference on a configurational interaction model as being caused by differential admixtures of $3d^84d4s$ states. Their theoretical evaluation of 126 MHz agreed well with the measured value of 141 MHz.

9.30. ZINC

Isotope shifts measured in $\lambda 589.4$ nm, the $3d^{10}4p\,^2P_{1/2}-3d^94s^2\,^2D_{3/2}$ transition of Zn II, are discussed by Foot *et al.* (1982). The shifts are predominantly mass shifts, but by comparison with the muonic work of Shera *et al.* (1976) and the electron scattering studies of Wohlfahrt *et al.* (1980) it was possible to determine the field shifts in the optical transition. The comparison was made with the aid of a King plot of the type shown in Fig. 6.6. The agreement between the muonic values of $\delta\langle r^2\rangle$, the changes between isotopes in the mean square radius of the nuclear charge distribution, and the optical isotope shifts was poor. It could only be said that there was agreement because of the large uncertainties in the muonic values of $\delta\langle r^2\rangle$. These large uncertainties arose mainly from allowing for the model dependence of the conversion of the muonic data into values of $\delta\langle r^2\rangle$. Foot *et al.* (1982) showed that if the known nuclear deformations are taken into account, as suggested by Stacey (1971), the agreement becomes somewhat better. The values of $\delta\langle r^2\rangle/\text{fm}^2$ that are consistent with both experiments are: [70,68]Zn, 0.15; [68,66]Zn, 0.13; [66,64]Zn, 0.14. For the isotope pair [66,64]Zn, the optical isotope shift was found to consist of a normal mass shift of 132 MHz, a field shift of 735 MHz, and a specific mass shift of -3710 MHz. The large value of the latter arises from the change in the number of 3d electrons during the transition, and had been predicted by Bauche (1969) to be about -4000 MHz.

9.31. GALLIUM

The isotope shift for [71,69]Ga has been measured in the resonance lines of Ga I $4s^24p\,^2P-4s^25s\,^2S_{1/2}$ (Neijzen and Dönszelmann, 1980). The sum of the specific mass shift and field shift was found to be -200 MHz for both transitions. Assuming that the heavier isotope has the larger nuclear charge distribution, a quite reasonable assumption in this region of the Periodic Table, the field shift in these transitions would be positive because of the 5s electron in the upper level. It then follows that the Vinti k factor is -0.14 a.u. if the field shift is negligible or more negative if the field shift is significant. Such k values are not inconsistent with the evaluated values of Bauche (1969) for neighboring elements, although no theoretical evaluations have been made for gallium. Jackson (1981) has commented on the comparison of the work of Neijzen and Dönszelmann with some earlier work of his own. This was his last paper on the subject of isotope shifts, and marks the end of a very long era; his first paper on the subject appeared forty nine years previously (Jackson, 1932).

9.32. GERMANIUM

The isotope shift has been measured in three lines of Ge I by Heilig *et al.* (1966). The transitions were all $4p^2-4p5s$, so the field shift would be expected to be positive. The isotope pair ^{76,74}Ge had the smallest field shift, and if this is assumed to be zero then the specific mass shift is -240 MHz, i.e., the Vinti k factor is -0.18 a.u. This is similar to the k factor in gallium for a 4p–5s transition. The field shifts for ^{74,72}Ge and ^{72,70}Ge were very similar and about 50 MHz larger than the field shift for ^{76,74}Ge.

9.33. ARSENIC

Arsenic has only one stable isotope, and there is nothing to report.

9.34. SELENIUM

Some isotope shift work was done at Hannover University (see Heilig, 1982), but has not yet been published in an accessible journal.

9.35. BROMINE

The magnetic dipole allowed transition $4s^2 4p^5\,^2P_{1/2}-4s^2 4p^5\,2P_{3/2}$ has been observed with a color center laser using magnetic rotation spectroscopy by Kasper *et al.* (1981). They found the ^{81,79}Br isotope shift to be 13.6^{+5}_{-2} MHz which is to be compared with a normal mass shift of 19 MHz.

9.36. KRYPTON

Many measurements of isotope shifts in Kr I have been reported, see Heilig (1977; 1982), the most recent was by Jackson (1980). The spectrum being complex, the size of the specific mass shift cannot be evaluated with confidence. This is also true for the field shift, since there is little information available about the nuclear charge distribution of the krypton isotopes. Gerhardt *et al.* (1979) quote a preliminary value obtained from muonic energies of $\delta\langle r^2\rangle = -0.03(2)$ fm^2 for ^{86,84}Kr. It is certain that the mass shifts are far larger than the field shifts in the lines which have been studied.

9.37. RUBIDIUM

The isotope shift for ^{87,85}Rb has been measured in the resonance lines $5s\,^2S_{1/2}-5p\,^2P_{3/2,1/2}$ by several workers. It was found that the isotope shift was significantly larger (in a positive direction) than the normal mass shift. The value of the specific mass shift has been calculated only according to the central-field model. This model has been shown to be inadequate in the case of potassium (Mårtensson and Salomonson, 1982), and is probably also inadequate for rubidium. The evidence is that the specific mass shift is small in the very similar transitions in Sr II (Bruch *et al.*, 1969), and negligible in this case (Thibault *et al.*, 1981a). It thus follows that the field shift in the transitions is positive and so the field shift of the 5s electron is positive and so the nuclear charge radius is smaller in ^{87}Rb than it is in ^{85}Rb. This is surprising but not unexpected since ^{87}Rb contains 50 neutrons, a magic number of neutrons. It is thus a very stable and compact nuclide compared with its neighbors. Its binding energy is also large compared with its neighboring nuclides. This relationship between isotope shift and binding energy in the vicinity of magic numbers of neutrons has been discussed by Gestenkorn, and is mentioned specifically for the case of rubidium by Brechignac *et al.* (1976).

Thibault *et al.* (1981a) studied the D_2 line of many radioactive isotopes which were produced either by the spallation of niobium or by the fission of uranium, both induced by 600-MeV protons from the CERN synchrocyclotron. The isotopes were transformed into a thermal atomic beam and allowed to interact with a perpendicular laser beam from a cw tunable dye laser. Optical resonance was detected through a magnetic deflection of the atoms. The measured isotope shifts showed a very marked discontinuity at ^{87}Rb with 50 neutrons. From ^{76}Rb to ^{87}Rb the mean square radius of the nuclear charge distribution tended to get smaller (as already known for ^{85}Rb and ^{87}Rb); from ^{87}Rb to ^{98}Rb it increased quite rapidly. The isotope shift for ^{97,96}Rb is particularly large at -350 MHz, the normal mass shift being 22 MHz and the specific mass shift probably smaller still. The large increase in $\langle r^2 \rangle$ from ^{96}Rb to ^{97}Rb is produced by a large rotational deformation in ^{97}Rb which has 60 neutrons; this is discussed in Section 10.2.

9.38. STRONTIUM

The isotope shifts between ^{90,88,86,84}Sr have been measured in λ460.7 nm $5s^2\,^1S_0-5s5p\,^1P_1$ of Sr I, and in the resonance lines $5s\,^2S_{1/2}-5p\,^2P$ of Sr II by Heilig (1961). With a hollow cathode light source, the detection system was sensitive enough for measurements to be made with small amounts of radioactive ^{90}Sr in the cathode. This was one of the earliest measurements of the isotope shift of a radioactive isotope. The shifts in the two Sr II resonance

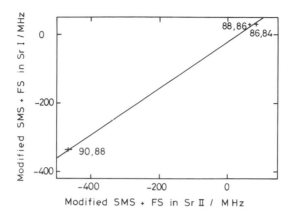

FIGURE 9.3. Optical isotope shifts in strontium. The shifts, modified to give the mass shift of $^{88,\,86}$Sr, in λ460.7 nm of Sr I ($5s^2$ $^1S_0$–5s5p $^1P_1$) are plotted against the mean of the modified shifts in the resonance lines of Sr II (5s $^2S_{1/2}$–5p $^2P_{1/2,\,3/2}$). The normal mass shifts have been subtracted from the measured shifts of Heilig (1961).

lines were very similar, and in Fig. 9.3 they are plotted against the shifts in λ460.7 nm of Sr I. In this plot the shifts have been modified to give the mass shifts for ^{88,86}Sr, and the normal mass shifts, of about 100 MHz, have been subtracted from the measured shifts. The field shift in these transitions is negative if the heavier isotope is larger, and so it is obvious from Fig. 9.3 that $\delta\langle r^2\rangle$ for ^{90,88}Sr is much larger than that for ^{88,86}Sr; the latter could even be negative, as it is for the same neutron numbers in rubidium. As in that element, there is an obvious discontinuity at the isotope with a magic number of neutrons, in this case ^{88}Sr.

The shifts plotted in Fig. 9.3 cannot be precisely divided into specific mass and field shifts since the former has not been calculated precisely and precise values of $\delta\langle r^2\rangle$ are not available from other experiments. It is clear from Fig. 9.3 that the specific mass shifts could be negligible as the plotted line could pass through the origin. On the assumption that the specific mass shifts in the Sr I line are very similar to those in the Sr II lines, it can be deduced that the specific mass shift is more likely to be negative than positive. It follows that the field shifts of ^{88,86}Rb, and ^{86,84}Rb are positive, and so $\delta\langle r^2\rangle$ is negative for these isotope pairs.

Ehrlich (1968) measured muonic *K*α isotope shifts and obtained a negative volume shift for ^{88,86}Sr, in agreement with the above analysis, and also for ^{87,86}Sr. This latter result agrees with the optical isotope shift of ^{87}Sr measured by Bruch *et al.* (1969) who found that the field shift for ^{87,86}Sr was about 0.9 of that for ^{88,86}Sr.

Since the spectral lines of Fig. 9.3 are simple transitions in simple spectra, it is very tempting to try to evaluate $\delta\langle r^2\rangle$ from the field shifts (obtained by assuming that the specific mass shifts are negligible), but the gradient of Fig. 9.3, which gives the ratio of the field shifts, counsels caution. For $5s^2-5s5p$ and $5s-5p$, the field shifts would be expected to be very similar. In fact, the gradient is actually nothing like unity but is 0.68. This low value shows that either Sr I or Sr II, if not both, is not such a simple spectrum as might be expected.

Beigang and Timmermann (1982) have measured isotope shifts in transitions involving Rydberg states of Sr I with the use of two-photon spectroscopy to overcome the Doppler broadening. As the principal quantum number varied from 10 to 70, the position of the ^{87}Sr isotope line moved dramatically with respect to ^{86}Sr and ^{88}Sr. This is not a pure isotope shift effect but is a spectacular example of the second order shifts which arise in alkaline earth spectra when the hyperfine and fine structure splittings are of comparable size; and effect already mentioned in the case of calcium and discussed in the case of Li II. Beigang and Timmermann (1982) show that the experimental results are in good agreement with a theoretical evaluation of the size of this hyperfine-induced singlet–triplet mixing. It affects the odd isotope, but not the even isotopes as these (even–even) nuclides have zero nuclear spin.

9.39. YTTRIUM

Yttrium has only one stable isotope, and there is nothing to report.

9.40. ZIRCONIUM

The isotope shifts between ^{96,94,92,90}Zr have been measured in visible and UV lines of Zr I by Heilig *et al.* (1963). The eleven lines could be divided into five types of transition and the residual shift ($SMS + FS$) of each type of transition is plotted against the mean of several shifts in Fig. 9.4. The Hartree–Fock calculations of Bauche (1974) for the case of molybdenum suggest that the specific mass shift will be small for transitions in which the number of 4d electrons does not change, about 600 MHz in a $4d^2-4d^3$ transition, and about -600 MHz in a $4d^3-4d^2$ transition for ^{92,90}Zr. The King plot of Fig. 9.4 is in agreement with this except that the $4d^35s\,a^5F_4-4d^35p\,y^3G$ transitions have fairly large negative specific mass shifts. These transitions are also anomalous in their field shifts, which are much smaller than those in the $4d^35s\,a^5F-4d^35p\,y^5G$ transitions. Otherwise the field shifts are as expected; they range from large negative values in $4d^25s^2-4d^35p$ to small positive values in $4d^35s-4d^25s5p$. The anomaly suggests the presence of configuration interaction. So far as the different isotope pairs are concerned, the field shift is largest for ^{92,90}Zr and smallest for ^{96,94}Zr; the values are given in Table 9.6. These results are completely

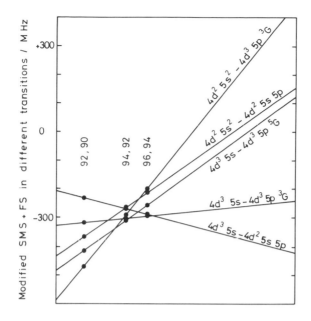

Mean of several modified isotope shifts

FIGURE 9.4. Optical isotope shifts in zirconium. The shifts, modified to give the mass shift of ^{92,90}Zr in various transitions, are plotted against the mean of several modified isotope shifts. The normal mass shifts have been subtracted from the measured shifts of Heilig *et al.* (1963).

incompatible with the results obtained by Fajardo *et al.* (1971) from electron zirconium scattering experiments, even allowing for the large uncertainties in the values they obtained for $\delta\langle r^2 \rangle$. Ehrlich (1968) measured muonic $K\alpha$ isotope shifts for ^{92,90}Zr and obtained $\delta\langle r^2 \rangle = 0.306(15)$ fm^2. This is larger than the value that would be obtained from the standard uniform model $R = 1.2\ A^{1/3}$ fm, as is to be expected since ^{90}Zr contains a magic number of neutrons.

9.41. NIOBIUM

Niobium has only one stable isotope, and there is nothing to report.

9.42. MOLYBDENUM

The isotope shifts for all the stable isotopes have been measured by Aufmuth *et al.* (1978) in sixteen lines of Mo I and six lines of Mo II by means of a photoelectric-recording Fabry–Pérot spectrometer. Allowance was made

TABLE 9.5

Isotope Shifts for ^{94,92}Mo in Mo I and Mo II

Transition between terms	Estimated specific mass shift (MHz)[a]	Estimated field shift (MHz)[a]
$d^5s\,^7S–d^5p\,^7P$	42	-663
$d^5s\,^5S–d^5p\,^5P$	36	-339
$d^5s\,^5S–d^5p\,^7P$	33	-381
$d^5s\,^7S–d^4sp\,^7P$	-288	72
$d^5s\,^5S–d^4sp\,^5P$	-483	489
$d^4s^2\,^5D–d^4sp\,^5P$	117	-684
$d^4s^2\,^5D–d^5p\,^5P$	636	-1512
$d^4s\,^6D–d^4p\,^6F$	42	-927
Fictitious transitions derived from above		
$d^5s\,^7S–d^5s\,^5S$	9	-282
$d^5p\,^7P–d^5p\,^5P$	-3	-42
$d^4sp\,^7P–d^4sp\,^5P$	186	-135

[a]The analysis was made by Aufmuth *et al.* (1978); their paper should be consulted for the uncertainties.

for configuration mixing with the aid of the work of Trees and Harvey (1952) so that the nine pure terms of Table 9.5 could be studied. The specific mass shifts given in Table 9.5 are a self-consistent set which (a) have small values except when the number of 4d electrons changes, (b) are consistent with muonic x-ray isotope shifts (Macagno *et al.*, 1970), and (c) agree qualitatively with Hartree–Fock calculations. The field shifts which result from these specific mass shift values are also given in Table 9.5. Two results of particular interest stand out: the large S dependence of (a) the specific mass shift and field shift in the $4d^45s5p$ configuration, (b) the field shift in the $4d^55s$ configuration. The former was unexpected but the latter was shown to be due to crossed second-order contributions between the field shift operator and the electrostatic operator (see Section 4.3).

The values of $\delta\langle r^2\rangle$ for various isotope pairs were obtained from the field shifts with the aid of the Goudsmit–Fermi–Segrè formula, Eq. (4.28), and Hartree–Fock calculations of the screening factors, Eq. (4.33), applied to $d^5s–d^5p$ and $d^4s–d^4p$ transitions. The results are given in Table 9.6. Schellenberg *et al.* (1980) have measured the energies of the $2p_{3/2}–1s_{1/2}$ and $2p_{1/2}–1s_{1/2}$ muonic x-ray transitions for various isotopes. Their $\delta\langle r^2\rangle$ values are inexplicably much smaller than those obtained from the optical work although the relative values are in reasonable agreement. The $\langle r^2\rangle$ values are model-dependent but the model error is most unlikely to be large enough to account for the discrepancy. They found $\langle r^2\rangle = 19.42(4)$ fm^2 for ^{98}Mo.

TABLE 9.6

Changes in the Mean Square Radius of the Nuclear Charge Distribution from Optical Isotope Shifts

Isotope pair by neutron number	$\delta\langle r^2\rangle(\text{fm}^2)$						
	$_{40}\text{Zr}^a$	$_{42}\text{Mo}$	$_{44}\text{Ru}^a$	$_{46}\text{Pd}^a$	$_{47}\text{Ag}$	$_{48}\text{Cd}$	$_{50}\text{Sn}$
Even, even							
52, 50	0.302(60)	0.226(19)					
54, 52	0.209(56)	0.193(16)	0.199(50)				
56, 54	0.163(52)	0.150(12)	0.176(44)				
58, 56		0.227(19)	0.162(41)	0.176(25)			
60, 58			0.178(45)	0.170(25)		0.152(19)	
62, 60				0.173(25)	0.144(9)	0.145(19)	
64, 62				0.153(25)		0.144(18)	0.129(6)
66, 64						0.130(17)	0.133(6)
68, 66						0.106(17)	0.124(6)
70, 68							0.112(5)
72, 70							0.101(5)
74, 72							0.091(5)
Odd, even							
53, 52		0.061(5)					
55, 54		0.027(2)	0.057(17)				
57, 56			0.036(14)				
59, 58				0.046(12)			
61, 60					0.022(3)		
63, 62					0.029(2)	0.031(19)	
65, 64						0.031(20)	0.047(6)
67, 66							0.049(6)
69, 68							0.045(5)

[a] Values of $\beta\delta\langle r^2\rangle$, where β is the screening factor of Eq. (4.33).

More accurate Doppler-free measurements of optical isotope shifts have been made by Siegel et al. (1981) who used a hollow cathode discharge tube with intermodulated fluorescence and optogalvanic spectroscopy. The shifts are consistent with those measured by Aufmuth et al. (1978).

9.43. TECHNETIUM

Technetium has no stable isotope, and there is nothing to report.

9.44. RUTHENIUM

No more recent isotope shift measurements are available than those of King (1964) who used a hollow cathode light source, a Fabry–Pérot interferometer, and photographic detection. Values of $\beta\delta\langle r^2\rangle$ were given by Heilig and Steudel (1974) and are included in Table 9.6. They should be treated with caution because of the experimental uncertainties in the measured shifts, and the large amount of configuration mixing shown to be present. Because of the latter, it is possible for the screening factor β to differ considerably from unity.

9.45. RHODIUM

Rhodium has only one stable isotope, and there is no work on isotope shifts to report. The model-dependent value $\langle r^2\rangle = 20.2(3)$ fm^2 was obtained by Engfer et al. (1974) from muonic x-ray studies of ^{103}Rh.

9.46. PALLADIUM

Isotope shifts have been measured by Baird (1976) who used a hollow cathode light source with digital recording interferometry. Values of $\beta\delta\langle r^2\rangle$ are given in Table 9.6; the relative values are accurate but the size of the specific mass shift, although assumed to be zero, is not known accurately and β may vary by up to about 15% from unity. No muonic x-ray energies have been measured.

9.47. SILVER

Isotope shifts have been measured by Fischer et al. (1975b) in the metastable isotopes ^{110}Ag and ^{108}Ag, as well as in the stable isotopes ^{109}Ag and ^{107}Ag. The former were produced by neutron irradiation in a reactor. The resonance lines $5s\,^2S_{1/2}-5p\,^2P_{3/2,1/2}$ from a hollow cathode light source were studied with Fabry–Pèrot etalons using photoelectric detection. There is good reason to think that the specific mass shifts are very small in such transitions, and the values of $\delta\langle r^2\rangle$ given in Table 9.6 were obtained on this assumption.

9.48. CADMIUM

Cadmium, with its eight stable isotopes, has long been a happy hunting ground for measurers of isotope shifts. Shifts have been measured in at least four optical transitions of Cd I and seven of Cd II as well as in the $K\alpha_1$ x-ray transition. A much studied optical transition is $\lambda441.6$ nm of Cd II, $4d^{10}5p-4d^95s^2$, because of the large field shifts produced by the change of 2 in the number of 5s electrons. Numerous measurements have also been made in the simpler transitions of Cd II, $4d^{10}5s\,^2S_{1/2}-4d^{10}5p\,^2P_{3/2,1/2}$. The specific mass shift is probably small in these lines, and the electronic factor of Eq. (4.24) and Eq. (4.33) can be estimated with reasonable confidence. The changes in the mean square radius of the nuclear charge distribution have been determined by Gillespie *et al.* (1975) from the isotope shift measurements of Brimicombe *et al.* (1976) and the results are given in Table 9.6.

The electronic x-ray isotope shifts measured by van Eijk *et al.* (1979) should, in principle, allow the determination of the optical specific mass shifts, but their shifts are not consistent with the optical isotope shifts. The nuclear charge radii determined from electron scattering experiments (Gillespie *et al.*, 1975) are barely consistent with the optical isotope shifts but they do give an absolute value for the mean square radius of the nuclear charge distributions. The results for ^{112,114,116}Cd are consistent, so absolute values of $\langle r^2 \rangle$ are probably best obtained by combining the figures of Table 9.6 with the electron scattering value for ^{114}Cd of $\langle r^2 \rangle = 21.43(7)$ fm^2 which was arrived at by assuming a two-parameter Fermi model for the charge distribution.

Wenz *et al.* (1981) also give values of $\delta \langle r^2 \rangle$, but they are based on relatively less accurate isotope shift measurements, even though they used Doppler-free saturation spectroscopy.

9.49. INDIUM

Isotope shifts have been measured for ^{115,113}In in the $5s^25p-5s^26s$ and $5s^25p-5s^27s$ transitions by Neijzen and Dönszelmann (1980) and by Eliel *et al.* (1981). The latter used a frequency-doubled frequency-stabilized cw ring dye laser of 2 MHz bandwidth to study the ultraviolet 5p–7s transition. The isotope shift is larger in this transition than in the 5p–6s transition. After subtraction of the normal mass shifts, the residual shifts, which should be predominantly field shifts for such transitions in a medium-heavy element, are 194 MHz for 5p–6s and 168 MHz for 5p–7s. The relatively large size of the latter suggests that the isotope shifts are not determined predominantly by the single outer electron, but by the change in the screening of the $5s^2$ electrons during the transition. It is thus not possible to estimate $\delta \langle r^2 \rangle$. The muonic x-ray work of Kast *et al.* (1971) shows that for ^{115}In, $\langle r^2 \rangle \simeq 4.62$ fm^2, a model-dependent value.

9.50. TIN

Isotope shifts of tin have been measured in muonic x-ray transitions by Macagno *et al.* (1970), in electronic x-ray transitions by Bhattacherjee *et al.*

(1969), and in optical transitions by many people, the most recently published work being that of Goble *et al.* (1974). With ten stable isotopes, many relative isotope shifts can easily be obtained, and variations with neutron number studied. It is more difficult to convert these relative shifts into changes in the mean square radius of the nuclear charge distribution because of uncertainties in the determination of the size of the specific mass shifts and the electronic factors of Eq. (4.26) and Eq. (4.32). Bishop and King (1971) reduced these difficulties by studying the particularly simple 5s–5p transitions of Sn IV. After allowance for the screening effects (King and Wilson, 1971) it was found that for ^{124,116}Sn $\delta\langle r^2 \rangle = 0.42(6)$ fm^2. The corresponding value obtained from electronic x-ray isotope shifts (Boehm and Lee, 1974) is 0.414(16) fm^2. Isotope shifts in muonic x-ray spectra (Macagno *et al.*, 1970) give $\delta\langle r^2 \rangle = 0.495(5)$ fm^2 for ^{124,116}Sn. The values of $\delta\langle r^2 \rangle$ for various pairs of isotopes which are given in Table 9.6 are taken from Goble *et al.* (1974).

9.51. ANTIMONY

Buchholz *et al.* (1978) measured isotope shifts in nine $5p^3-5p^2 6s$ transitions. The specific mass shifts were evaluated by Bauche (1974) to be about -120 MHz from which it followed that the field shifts were about 160 MHz, and the change in the mean square radius of the nuclear charge distribution for ^{123,121}Sb was 0.12(4) fm^2.

9.52. TELLURIUM

Little isotope shift work has been done on this element whose complicated spectrum is difficult to excite. Four lines of Te II have been studied by Lecordier and Helbert (1978) and Lecordier (1979). The residual shifts of the spectral lines are plotted against each other in Fig. 9.5. The size of the specific mass shifts is difficult to evaluate in such a complicated spectrum but it is clear from Fig. 9.5 that the field shifts for $\delta N = 2$ between even isotopes decrease steadily as A increases. It is also clear that there is considerable odd–even staggering; exactly how much depends on the size of the specific mass shift. Two estimates of this have been made. Firstly, according to a Hartree–Fock evaluation the *SMS* in the various spectral lines are given by a vertical line through A in Fig. 9.5. Secondly, use has been made of the correlation between relative field shifts and variations of the quantity $EA^{1/3}$, where E is the binding energy per nucleon, which has been pointed out by Gerstenkorn (1971). If the size of specific shift is chosen which maximizes this correlation for the even isotopes then the specific mass shifts are given by a vertical line through B in Fig. 9.5. With such a large disparity between the

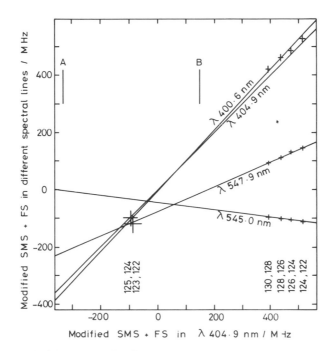

FIGURE 9.5. Optical isotope shifts in tellurium. The shifts, with the normal mass shifts removed, have been modified to give the mass shifts of ^{130,128}Te. Results for the four lines measured (Lecordier and Helbert, 1978; Lecordier, 1979) are plotted against the shifts in $\lambda 404.9$ nm. The latter line, plotted against itself, is shown without experimental points. Vertical lines through A and B give possible sets of values for the specific mass shift mentioned in the text.

two estimates, it is not possible to deduce reliable values of $\delta \langle r^2 \rangle$ between the tellurium isotopes.

9.53. IODINE

Iodine has only one stable isotope, ^{127}I, but the possibility of removing ^{127}I from enriched ^{129}I (half-life 1.7×10^7 years) by a laser process led Engleman *et al.* (1980) to measure the ^{129,127}I isotope shift in the $\lambda 1.3$ μm transition with the aid of high-resolution Fourier transform spectroscopy. They concluded, from the small size of the isotope shift, that laser separation would be difficult at best. Wilson (1968) made *ab initio* Hartree–Fock calculations of the electron density at the nucleus for many electron configurations, but this was done because of interest in Mössbauer isomer shifts, not isotope shifts.

9.54. XENON

Isotope shifts have been measured in a very large number of optical transitions, and Jackson *et al.* (1975) made an extremely thorough parametric study of 30 levels of Xe I. King plots of numerous transitions gave the relations between the mass shifts and field shifts and these were found to be in satisfactory agreement with the parametric study which used six parameters. The results were also in reasonable agreement with *ab initio* Hartree–Fock evaluations. Picking out one transition in particular for comparison with other work, Jackson *et al.* (1975) found that in λ979.9 nm, $5p^5(^2P_{3/2})6s[\frac{3}{2}]2 - 5p^5(^2P_{3/2})6p[\frac{1}{2}]1$, for ^{136,134}Xe, $IS = -106(10)$ MHz, $NMS = 18$ MHz, $SMS = -12$ MHz, and $FS = -112$ MHz.

Fischer *et al.* (1974b) measured isotope shifts in four lines of Xe I including λ979.9 nm. They deduced the size of the specific mass shift by comparing isotope shifts in Xe with isotope shifts in Ba. The variations in $\delta\langle r^2 \rangle$ between isotope pairs are known to be fairly similar for neighboring elements in the Periodic Table if the comparison is made for the same pair of neutron numbers. It thus follows that a King plot of the shifts in a xenon transition against the shifts in a barium transition will give a reasonable approximation to a straight line if the same pairs of neutron numbers are used for both elements at each point. Fischer *et al.* (1974b) found such a King plot gave a good enough straight line for the specific mass shift in the Xe I transition to be deduced from the relatively better known specific mass shift in the Ba II transition. The shifts they obtained in λ979.9 nm for ^{136,134}Xe were, $IS = -102(2)$ MHz, $NMS = 18$ MHz, $SMS = -39$ MHz, and $FS = -81$ MHz.

These two estimates of the specific mass shift obviously differ by more than the experimental uncertainty in the measurement of the isotope shifts. Neither method claims to give exactly the correct value and it is probably best to take a mean. (AUTHOR's NOTE: I feel that the parametric study, since it gives self-consistent results for many transitions, should be given more weight and so I adopt the value of -20 MHz for the *SMS* in λ979.9 nm for ^{136,134}Xe and estimate the uncertainty as about ± 20 MHz.)

Fischer *et al.* (1974b) used the Goudsmit–Fermi–Segrè formula to obtain $\pi|\psi(0)|^2 a_0^3/Z = -0.162$, which, with Wilson's evaluation of the screening factor $\beta = 1.16$ and $f(Z) = 11.37$ GHz/fm^2 (Table 4.2), gives the relation between field shift in MHz, and change in nuclear charge distribution in fm^2,

$$\delta\langle r^2 \rangle = -4.68 \times 10^{-4} \, FS_{\lambda979.9} \qquad (9.2)$$

Values for various isotope pairs based on the above analysis are given in Table 9.7 which also includes the measurements of Hühnermann *et al.* (1978)

TABLE 9.7
Isotope Shifts and Changes of Nuclear Radii in Xenon

Isotope pair	Isotope shift (MHz)[a]	Field shift (MHz)[b]	$\delta\langle r^2\rangle$(fm^2)[c]
126,124	−89(2)	−87	0.041
128,126	−80(2)	−78	0.036
130,128	−69(2)	−67	0.031
132,130	−50(3)	−48	0.022
134,132	−66(3)	−64	0.030
136,134	−102(2)	−100	0.047
129,128	+9(2)	+10	−0.005
132,131	−87(2)	−86	0.040
134,133	−99(6)	−98	0.046

[a] From measurements by Fischer *et al.* (1974b) and Hühnermann *et al.* (1978) in λ979.9 nm.
[b] Because of the uncertainty in the mass shift, every figure in this column could be in error by the same amount, which amount could be anything from −20 to +20.
[c] The uncertainty in the conversion factor from field shift to $\delta\langle r^2\rangle$ is about 10%.

on the radioactive ^{133}Xe isotope. They collected a few ng of ^{133}Xe in a hollow cathode by means of the Marburg mass separator, and with this were able to study three infrared lines, including λ979.9 nm.

More recently there have been two laser spectroscopic measurements of isotope shifts in Xe II (Alvarez *et al.*, 1979; Borghs *et al.*, 1981), but they do not help to reduce the uncertainties of Table 9.7.

9.55. CESIUM

The simple spectrum of cesium, an alkali metal, has encouraged spectroscopists to measure optical isotope shifts even though there is only one stable isotope, ^{133}Cs. Hühnermann and Wagner (1967) studied the 6s–6p transitions of radioactive ^{135}Cs and ^{137}Cs using a hollow cathode light source with a Fabry–Pérot interferometer and photoelectric recording. Gerhardt *et al.* (1978) studied the 6s–7p transitions of the same isotopes by laser saturation spectroscopy. Figures that can be deduced from their results are given in Table 9.8. These show that the field shifts are the same size in all four transitions within experimental uncertainties. It can be seen from a King plot that the specific mass shifts are also very similar in the four transitions.

According to Fradkin (1962)—see Section 4.4—the field shift of $6^2S_{1/2}$–$6^2P_{1/2}$ should be smaller than that of $6^2S_{1/2}$–$6^2P_{3/2}$ by 4.4% if the field shift is predominantly due to the 6s electron as it presumably is in this case. The results in the last column of Table 9.8 are in complete agreement with this within the experimental uncertainties. More accurate measurements might however reveal interesting effects since, as Gerhardt *et al.* (1978) point

TABLE 9.8
Isotope Shifts in Cs I[a]

Transition	SMS + FS (MHz)		
	135,133	137,135 (modified)	Difference
$6^2S_{1/2}-6^2P_{1/2}$	$-57.4(2.1)$	$-130.7(2.5)$	$-73.3(3.2)$
$6^2S_{1/2}-6^2P_{3/2}$	$-57.8(4.5)$	$-134.5(2.5)$	$-76.7(5.1)$
$6^2S_{1/2}-7^2P_{1/2}$	$-57.9(1.0)$	$-132.4(1.0)$	$-74.5(1.4)$
$6^2S_{1/2}-7^2P_{3/2}$	$-58.7(1.0)$	$-132.6(1.0)$	$-73.9(1.4)$

[a]The shifts for ^{137,135}Cs have been modified to give the mass shifts for ^{135,133}Cs. The normal mass shifts have been subtracted from the measured shifts. The difference column shows the relative sizes of the field shifts in the four transitions.

out, their hfs measurements, as well as those of others, show that the 6p and 7p configurations are not pure but interact with each other.

As cesium has only one stable isotope there are no muonic *IS* measurements from which to deduce the size of the *SMS*s in the optical transitions. All that is certain is that they are very similar in all four transitions, but in the light of the evidence from other alkali-like spectra it is reasonable to assume that the specific mass shift is comparable with or smaller than the normal mass shift in the 6s–6p transition which is about 21 MHz. Thibault *et al.* (1981b) have measured the isotope shift in the $6^2S_{1/2}-6^2P_{3/2}$ transition for a vast range of radioactive isotopes stretching from ^{118}Cs to ^{145}Cs. These were produced by spallation of lanthanum or by fission of uranium by a 600-MeV proton beam at CERN. The isotopes were turned into a thermal atomic beam and made to interact with a tunable dye laser. To convert their isotope shifts into values of $\delta\langle r^2 \rangle$ they assumed that the specific mass shift was zero. The screening factor β of Eq. (4.33) was taken to be 1.1 and $|\psi(0)|^2$6s was found from the magnetic hyperfine splitting factor of the 6s electron. The relation between field shift and $\delta\langle r^2 \rangle$ was found to be -2313 MHz/fm^2, and the values obtained for $\delta\langle r^2 \rangle$ by Thibault *et al.* (1981b) are shown in Fig. 9.6. An absolute value for the mean square radius of the nuclear charge distribution of ^{133}Cs has been obtained by Engfer *et al.* (1974) from muonic x-ray energy measurements, $\langle r^2 \rangle = 23.10$ fm^2, and this has been incorporated into Fig. 9.6. If the *SMS* were not zero then the polygon of Fig. 9.6 would be tipped about the ^{133}Cs point but with little change in its shape. For example, if the *SMS* were 10 MHz for ^{135,133}Cs, then $\langle r^2 \rangle$ of ^{118}Cs and ^{119}Cs would be reduced by 0.037 fm^2 and 0.034 fm^2, respectively, and $\langle r^2 \rangle$ of ^{144}Cs and ^{145}Cs would be increased by 0.022 fm^2 and 0.024 fm^2, respectively. The very obvious discontinuity at ^{133}Cs—the isotope with a magic number of neutrons—remains whatever the size of the *SMS*. Other features of Fig. 9.6 are discussed in Section 10.2.

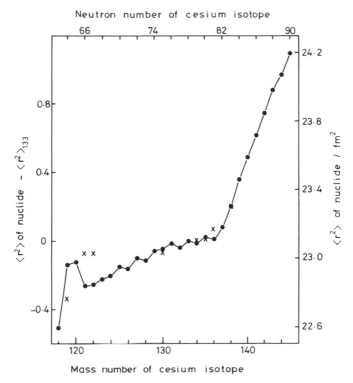

FIGURE 9.6. Mean square radii of the nuclear charge distribution of cesium isotopes. Points labeled (●) are ground state nuclides; points labeled (×) are isomers. Data from Thibault *et al.* (1981b) and Engfer *et al.* (1974).

9.56. BARIUM

Barium has five common stable isotopes. Their isotope shifts have been measured in optical spectra (Baird *et al.* 1979) and their muonic x-ray transition energies have been measured in various transitions (Shera *et al.*, 1982). Optical isotope shifts have been measured between these isotopes and many unstable neutron deficient isotopes by Bekk *et al.* (1979), who were also able to study three isomers ($A = 129, 133, 135$). These isotopes were produced as beams via compound nuclear reactions by charged particle irradiation of appropriate targets using deuteron and α-particle beams of a cyclotron. The optical transition used was $\lambda 553.5$ nm of Ba I ($6s^2\,^1S_0 - 6s6p\,^1P_1$). Fischer *et al.* (1974a) measured shifts in $\lambda 493.4$ nm of Ba II, the alkali-like resonance transition $6s\,^2S_{1/2} - 6p\,^2P_{3/2}$, and included radioactive ^{140}Ba chemically separated from fission products of uranium. There are several interconnected ways of analyzing this data to obtain values of $\delta\langle r^2 \rangle$ between barium nuclides. These ways are displayed in Fig. 9.7 and will now be discussed.

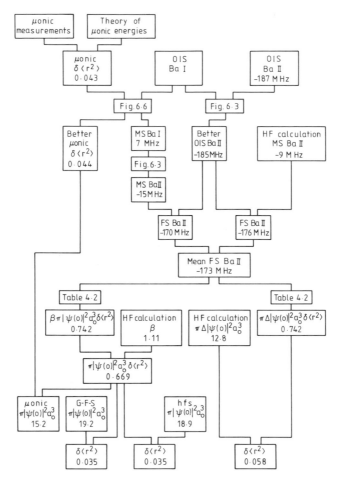

FIGURE 9.7. Methods of determining $\delta\langle r^2\rangle$ for barium isotopes. The diagram displays the various possible methods. The values of $\delta\langle r^2\rangle$ are given in fm^2 and refer to the isotope pair ^{138,136}Ba. The following abbreviations are used: *OIS*, optical isotope shift; *MS*, mass shift; *FS*, field shift; Ba I, the $\lambda553.5$-nm transition $6s^2\,^1S_0-6s6p\,^1P_1$; Ba II, the $\lambda493.4$-nm transition $6s\,^2S_{1/2}-6p\,^2P_{3/2}$.

Fischer *et al.* (1974a) calculated the mass shift in the simple transition $\lambda493.4$ nm to be -9 MHz for ^{138,136}Ba. It follows from Fig. 6.3 that the corresponding mass shift in $\lambda553.5$ nm is 12 MHz. Comparison of shifts in this spectral line with $\delta\langle r^2\rangle$ derived from muonic data (Fig. 6.6) gives a mass shift in $\lambda553.5$ nm of 7(11) MHz. Combining these two independent results of comparable uncertainty gives a mass shift in $\lambda553.5$ nm of 9(9) MHz for ^{138,136}Ba.

Taking the mass shift as 9 MHz, the best fit of Fig. 6.6 can be recalculated with an extra point at 9 MHz, 0 fm^2. The slope of this best-fit line is 0.00032(1) fm^2/MHz and the corresponding values of $\delta\langle r^2\rangle$ can be calculated for all the isotopes and isomers whose shifts have been measured in $\lambda 553.5$ nm. These results are shown in Fig. 2.2. The shape of this polygon is precisely known because the optical isotope shift measurements were made with great accuracy using laser spectroscopic techniques. The absolute position of the polygon is less well known, because this depends on the value of the mass shift in $\lambda 553.5$ nm and the conversion factor from field shift to $\delta\langle r^2\rangle$. Taking the mass shift as 0 or 18 MHz, a best-fit to the muonic data has a slope of 0.00034(1) or 0.00030(1) fm^2/MHz, respectively. The corresponding values of $\delta\langle r^2\rangle$ are given in Table 9.9 and they have been indicated, for the extreme nuclides only, in Fig. 2.2. The effect of increasing the mass shift is to lower the points for nuclides in Fig. 2.2 away from ^{138}Ba. It can be seen from Table 9.9 that the effect on ^{137}Ba, ^{136}Ba, ^{135}Ba, ^{134}Ba, and ^{133}Ba is very small. Odd–even staggering and the general shape of the polygon formed by the points in Fig. 2.2 are affected only very slightly by the value chosen for the mass shift in $\lambda 553.5$ nm. Shera $et\ al.$ (1982) measured muonic x-ray transition energies in more than one transition and so were able to calculate $\langle r^2\rangle$ for each isotope they studied. Their value for $\langle r^2\rangle$ of ^{138}Ba, 23.348(3) fm^2, has been chosen (arbitrarily) for the scale on the right hand side of Fig. 2.2.

TABLE 9.9
Mean Square Radii of Charge Distribution of Barium Nuclides Relative to
That of ^{138}Ba(fm^2)

Mass number	Assumption		
	(a)	(b)	(c)
140	+0.285(f)	+0.299	+0.313(f)
137	−0.068	−0.070	−0.073
136	−0.044	−0.044	−0.043
135	−0.087	−0.088	−0.088
134	−0.054	−0.052	−0.048
133	−0.090	−0.088	−0.085
132	−0.068	−0.063	−0.057
131	−0.095	−0.090	−0.084
130	−0.086	−0.079	−0.070
129	−0.121	−0.114	−0.106
128	−0.111(f)	−0.002	−0.092(f)
126	−0.143(f)	−0.133	−0.121(f)
135(m)	−0.057	−0.056	−0.055
133(m)	−0.075	−0.072	−0.068
129(m)	−0.136	−0.130	−0.123

[a]The figures are obtained by combining nuclear size data from muonic atom spectra with optical isotope and isomer (m) shifts measured in $\lambda 553.5$ nm. The mass shift of ^{138,136}Ba in $\lambda 553.5$ nm is assumed to be (a) 18 MHz, (b) 9 MHz, and (c) 0. The corresponding best fit with the muonic data gives the ratio of $\delta\langle r^2\rangle$ to field shift as (a) 0.00030 fm^2/MHz, (b) 0.00032 fm^2/MHz, and (c) 0.00034 fm^2/MHz. The figures of column (b) and those marked (f) are plotted in Fig. 2.2.

Values of $\delta\langle r^2\rangle$ can also be obtained by the use of optical isotope shifts only. Methods of doing this are outlined in Fig. 9.7 and have already been mentioned in Section 4.2.1. The Hartree–Fock calculation of

$$\beta = \frac{\Delta|\psi(0)|^2(5p^66s-5p^66p)}{|\psi(0)|^2(6s\ in\ 5p^66s)} = 1.11 \tag{9.3}$$

gives the same value for β as does the pseudorelativistic calculation (Wilson, 1978a). This suggests that the value is reasonably accurate and so can be combined with empirically determined values of $|\psi(0)|^2(6s)$ to give an accurate value of $\delta\langle r^2\rangle$. Using the Goudsmit–Fermi–Segrè formula or the known hfs splittings gives $\delta\langle r^2\rangle = 0.035$ fm^2 for ^{138,136}Ba. The corresponding value obtained from muonic energies is 0.044 fm^2 (see Table 9.9), so if the muonic work is to be ignored all the values of $\delta\langle r^2\rangle$ in Table 9.9 must be multiplied by $0.035/0.044 = 0.80$. The discrepancy between the muonic and optical work is thus 20%. It is difficult to decide how much weight to give to each, so the best estimates of $\delta\langle r^2\rangle$ are probably to be made by giving them equal weight, and hence multiplying the figures in column (b) of Table 9.9 by 0.9. The conclusions to be drawn from these results so far as nuclear structure are concerned are discussed in Section 10.2.

Returning to the electronic structure of barium, Jitschin and Meisel (1980) have measured isotope shifts in many optical spectral lines. They used a Doppler-free two-photon absorption method which enabled them to reach many high-lying levels from the ground state $6s^2\,^1S_0$. The field shifts of these various transitions can be compared by means of King plots. The ratio of field shifts is equal to the ratio of $\Delta|\psi(0)|^2$, the change in the total electron density at the nucleus during the transition. The values of $\Delta|\psi(0)|^2$ relative to that for the transition $6s^2\,^1S_0-6s6p\,^1P_1$ ($\lambda553.5$ nm) were determined by Jitschin and Meisel (1980) and are given in Table 9.10. It can be seen that the electron density at the nucleus is not constant within different levels of a term that nominally belong to the same configuration. This indicates strong second-order mixing effects in the upper states. Taking the mass shift as 9 MHz for ^{138,136}Ba in $\lambda553.5$ nm, the mass shift in the other transitions can be found. The specific mass shifts are fairly similar in all the transitions except the two that have a 5d electron in the upper level, these two being very similar to each other. The figures are given in Table 9.10.

Grundevik *et al.* (1982) have measured isotope shifts in a number of 6s5d–6p5d transitions. One transition, $^1D_2-^1P_1$, was found to be unique; it had a negligible field shift, and so its specific mass shift could be determined unambiguously to be $-55(3)$ MHz for ^{138,136}Ba. They interpreted the lack of a field shift as indicating a strong rearrangement of the core electrons during the transition, rather than the presence of configuration mixing.

TABLE 9.10
Relative Field Shifts and Mass Shifts in Ba I[a]

Upper level of transition	Relative $\Delta\lvert\psi(0)\rvert^2$	Mass shift in ^{138,136}Ba (MHz)	Specific mass shift in ^{138,136}Ba (MHz)
$6s6p\,^1P_1$	1.00	9	-22
$6p^2\,^3P_0$	1.54(3)	14(4)	$-47(4)$
$6p^2\,^3P_2$	1.08(3)	34(4)	$-29(4)$
$6s8s\,^1S_0$	0.80(2)	32(2)	$-28(2)$
$6s9s\,^1S_0$	0.59(1)	46(3)	$-20(3)$
$6s7d\,^1D_2$	1.52(4)	28(8)	$-34(8)$
$6s7d\,^3D_2$	0.67(1)	41(2)	$-22(2)$
$6s8d\,^1D_2$	0.94(1)	38(2)	$-28(2)$
$5d7s\,^1D_2$	1.63(2)	105(2)	46(2)
$5d6d\,^3D_2$	1.94(2)	121(2)	58(2)

[a]The figures are derived from Jitshin and Meisel (1980), but the mass shifts are based on the values in the first line and so differ from theirs. The uncertainties in the mass shifts are based on the assumption that there is no error in the mass shifts given in the first line. All the transitions share the same lower level, $6s^2\,^1S_0$.

Neukammer and Rinneberg (1982) have studied the $6sns$ Rydberg levels of Ba I for $14 \le n \le 23$. The $n = 15$, 19, and 20 levels were found to be perturbed. The field shifts between ^{138,136}Ba and ^{136,134}Ba of $6s6p\,^1P_1-6sns\,^3S_1$ were compared and found to be very similar except for the perturbed levels. Even more sharply defined discontinuities showed up in the hyperfine splitting factor A, and the authors preferred to use this to determine the amount of admixture of the perturber levels into the Rydberg levels. The isotope shifts involving odd isotopes have also been measured in Rydberg levels (Rinneberg *et al.*, 1982) where again perturbation was found. The $6s18s\,^1S_0$ level is perturbed by $5d7d\,^3P_0$ in the case of the even isotope. In the case of the odd isotopes, the magnetic hyperfine interaction with the nearby $6s18s\,^3S_1$ level is also significant, and introduces hyperfine-induced isotope shifts as already mentioned in Section 9.3.1 for the case of lithium. The amount of perturbation is sensitive to energy changes of the order of the magnetic hyperfine splitting and the isotope shift, and so it varies from one isotope to another. These magnetic hyperfine interactions, varying from isotope to isotope, lead to a nonlinearity when the isotope shifts in $6s6p\,^1p_1-6s18s\,^1S_0$ are plotted against the shifts in $6s^2\,^1S_0-6s6p\,^1P_1$. Rinneberg *et al.* (1982) show that the linearity is restored when allowance is made for the magnetic hyperfine interaction between the three levels. The amount of admixture required to give linearity is consistent with the amount needed to bring the fine structure separations into agreement with the Landé interval rule.

The simple alkali-like spectrum of Ba$^+$ has attracted spectroscopists for a long time, and several isotope shifts have been measured in Ba II. The work of Fischer *et al.* (1974a), from which mass shifts in Ba I lines were deduced, has been mentioned already. More recently, Höhle *et al.* (1978) have measured isotope shifts in $5d\,^2D_{3/2}-6p\,^2P_{3/2}$. Metastable barium ions in a fast ion beam were excited by laser light, and the subsequent fluorescence light was observed. Due to the different velocities of the different isotopes, the contributions of the various isotopes to the fluorescence signal were well separated. The field shift in the 5d–6p transition was found to be 0.22 of that in the 6s–6p transition (but in the opposite direction), a fraction which was thought to be surprisingly large at the time. Wilson (1978) made some *ab initio* calculations and showed that the screening of 5s by 5d is indeed considerable. His pseudorelativistic evaluation of the above ratio was 0.23, a figure in very good agreement with the experimental value. The details are given in Table 9.11.

Wilson (1978a) also calculated the specific mass shift in 6p–5d to be -144 MHz for ^{138,136}Ba, which is in good agreement with the value obtained by Höhle *et al.* (1978).

Even more recently the isotope shifts in $5d\,^2D_{3/2}-6p\,^2P_{3/2}$ and $5d\,^2D_{5/2}-6p\,^2P_{3/2}$ have been measured by Van Hove *et al.* (1982). They used fast-ion-beam laser spectroscopy in a collinear geometry and achieved sufficient precision to demonstrate a (J-dependent) difference in the field shifts of the two transitions. The field shift of $D_{3/2}-P_{3/2}$ was 1.0465(10) times that of $D_{5/2}-P_{3/2}$. Using the calculated value of the specific mass shift in these transitions (-144 MHz) it can be deduced that the mass shift in $\lambda553.5$ nm of Ba I is 2 MHz or -3 MHz. These values are in good agreement with the previously used estimate of 9 MHz. The agreement is perhaps better appreciated when the argument is reversed. If the mass shift in $\lambda553.5$ nm is 9 MHz,

TABLE 9.11
Pseudorelativistic Values of $4\pi|\psi(0)|^2a_0^3$ for Ba II[a]

Contributing electrons	Valence electron		
	6s	6p	5d
$1s^2$	2,861,691	2,861,690	2,861,695
$2s^2$	323,024	323,023	323,024
$3s^2$	65,180	65,177	65,175
$4s^2$	14,049	14,047	14,043
$5s^2$	2,168	2,163	2,133
6s	115		
Total	$\overline{3,266,228}$	$\overline{3,266,100}$	$\overline{3,266,071}$

$$\frac{\Delta|\psi(0)|^2(5d\text{-}6p)}{\Delta|\psi(0)|^2(6s\text{-}6p)} = \frac{-29}{128} = -0.23$$

[a]The figures are from Wilson (1978a).

then the specific mass shifts are -146 MHz in $D_{3/2}-P_{3/2}$ and -147 MHz in $D_{5/2}-P_{3/2}$. In fact, all the recent work on isotope shifts in Ba II has confirmed the estimate of the mass shift in $\lambda 553.5$ nm of Ba I from which, in combination with muonic data, the nuclear charge radii of Table 9.9 and Fig. 2.2 were deduced.

9.57. LANTHANUM

Fischer *et al.* (1974c) have measured optical isotope shifts in five lines of La I between ^{137}La, ^{138}La, and ^{139}La, Childs and Goodman (1979) in two lines between ^{138}La and ^{139}La only. The spectrum is complex, configuration mixing is rife, and little more can be said than that the specific mass shift is significant (k factors up to 0.4 a.u. in the transitions), and that the nuclide with 81 neutrons is smaller than the one with 80 neutrons (as is also the case for neighboring barium; see Fig. 2.2).

9.58. CERIUM

To avoid the complexities of a lanthanide spectrum, Edwin and King (1969) measured isotope shifts in the resonance lines of the alkali-like Ce IV. They found that $\beta\delta\langle r^2\rangle$ for ^{142,140}Ce was 0.290(12) fm^2, which together with a Hartree–Fock evaluation of the screening factor (King and Wilson, 1971) gave $\delta\langle r^2\rangle = 0.265(12)$ fm^2. Isotope shifts have also been measured in the $K\alpha_1$ x-ray transition by van Eijk and Visscher (1970) and they deduced that, for ^{142,140}Ce, $\delta\langle r^2\rangle = 0.274(10)$ fm^2. These results are in a very good agreement and have very similar uncertainties, but the reasons for the two uncertainties are quite different. In the optical case the uncertainty arises mainly from a lack of knowledge of the specific mass shift. The uncertainty in this quantity was put, somewhat arbitrarily, at ±90 MHz, which is to be compared with a field shift of about 3000 MHz. In the x-ray case the mass shift can be evaluated and the uncertainty arises from experimental difficulties in measuring the isotope shift.

Isotope shifts have also been measured in the more complex Ce I and Ce II by Champeau (1972), Fischer *et al.* (1975a), and Champeau and Verges (1976), and the results obtained are more or less explicable and in reasonable agreement with *ab initio* calculations (Wilson, 1978b). A King plot (Fig. 6.4) of some of these transitions has already been mentioned in Section 6.1 because of the interesting case of $\lambda 2399.9$ nm. This line has a very small field shift and so its specific mass shift is known fairly precisely to be 640(30) MHz. It is thought to be a fairly pure $4f5d^2-4f^26s$ transition (Champeau and Verges, 1976) and so, since the predominant specific mass shift in such a transition is

TABLE 9.12

Field Shifts and Differences of Nuclear Charge Radii between Isotopes of Cerium

Isotope pair	138,136	140,138	140,136	142,140	144,142
Field shift (MHz) in:					
λ1708.0 nm			−204(80)	−3829(50)	
λ825.3 nm	71(38)	−245(37)	−174(70)	−3279(34)	
λ529.1 nm				−2405(29)	−2071(95)
$\delta\langle r^2\rangle$(fm)	−0.006	0.020		$0.270(10)^a$	0.233

[a] Average value from King and Wilson (1971) and van Eijk and Visscher (1970). The uncertainty in $\delta\langle r^2\rangle$ for another isotope pair involves the uncertainty in this value combined with the uncertainties in the field shifts which, except for ^{144,142}Ce, are predominantly the uncertainty in the mass shift of the relevant transitions.

due to the change in the number of 4f electrons, the specific mass shift of such a change is known. λ1708.0 nm is another transition between two pure configurations, this time 4f5d6s–4f ^{2}5d, and so its specific mass shift should be very similar as it again involves 4f–4f ^{2}. Wilson (1978b) has carried out HFR evaluations of the specific mass shifts which suggest that the specific mass shift in 4f5d6s–4f ^{2}5d is larger than that in 4f5d^2–4f ^{2}6s by about 150 MHz. So taking the specific mass shift in λ1708.0 nm to be 790(40) MHz the field shifts are −3829(50) MHz for ^{142,140}Ce, and −204(80) MHz for ^{140,136}Ce. A comparison with shifts measured for other isotopes in λ825.3 nm and λ529.1 nm of Ce I (Champeau, 1972; Fischer et al., 1975a) enables the relative field shifts of Table 9.12 to be obtained. These have been combined with the average value of $\delta\langle r^2\rangle = 0.270$ fm^2 for ^{142,140}Ce to give values of $\delta\langle r^2\rangle$ for the other isotope pairs.

The pseudorelativistic Hartree–Fock calculations (HFR) of Wilson (1978b) have been used in preference to those obtained using Hartree–Fock (HF), wavefunctions because their superiority shows up in the evaluation of the relative sizes of the field shifts in transitions between different electron configurations. The comparisons with the experimental results of Champeau

TABLE 9.13

Relative Field Shifts in Ce II

Transition	Hartree–Fock evaluation[a]	HFR evaluation[b]	Experimental result[a]
4f ^{2}6s–4f ^{2}6p	1.00	1.00	1.00
4f5d^2–4f ^{2}6s	0.43	0.12	−0.03
4f5d^2–4f ^{2}5d	1.67	1.34	1.12
4f5d6s–4f ^{2}5d	3.21	2.74	2.40

[a] From Champeau and Verges (1976).
[b] From Wilson (1978b).

and Verges (1976) are given in Table 9.13. The pseudorelativistic approach was suggested by Cowan and Griffin (1976) and is discussed in Section 4.2.2 where it is applied to isotope shifts in Ba II.

9.59. PRASEODYMIUM

Praseodymium has only one stable isotope, and there is nothing to report.

9.60. NEODYMIUM

Isotope shifts have been measured in several lines of Nd II (King *et al.*, 1973) and Nd I (van Leeuwen *et al.*, 1981). They have also been measured in electron $K\alpha$ x-ray transitions (Lee and Boehm, 1973) and muonic x-ray transitions (Macagno *et al.*, 1970). The comparison of x-ray with optical work to deduce values of $\delta \langle r^2 \rangle$ is discussed by King *et al.* (1973). From ^{142}Nd to ^{150}Nd, the neutron number changes from 82 (a magic number) to 90 and the deformation increases considerably. The lighter isotopes have a vibrational deformation, whereas ^{150}Nd has a permanent deformation and behaves like a rigid rotator. The amount of deformation is known from Coulomb excitation experiments and so the $\delta \langle r^2 \rangle$ values can be split into shape and volume components by the use of Eq. (4.42). This has been done in Table 9.14. On the electronic side, it is difficult to make precise deductions because of the complicated nature of Nd I and Nd II with their large amounts of configuration interaction. The work of van Leeuwen *et al.* (1981) was of sufficient

TABLE 9.14
Deformations and Radii of Neodymium Nuclei

Isotope	β_2^2	$\langle r^2 \rangle (\text{fm}^2)^a$	$\delta \langle r^2 \rangle (\text{fm}^2)^b$	$\delta \langle r^2 \rangle_{\text{shape}} (\text{fm}^2)^c$
142	0.0108	24.10		
			0.277	0.04
144	0.0151	24.38		
			0.257	0.08
146	0.0228	24.62		
			0.286	0.16
148	0.0388	24.92		
			0.381	0.38
150	0.0778	25.30		

[a] Electron scattering values adjusted to be consistent with the $\delta \langle r^2 \rangle$ values from optical isotope shifts.
[b] From King *et al.* (1973).
[c] Using Eq. (4.42).

accuracy to show up the hyperfine perturbation of the isotope shifts of the odd isotopes as discussed in Section 9.3.1 for the case of Li II. As already mentioned, the neodymium spectra are very complex and their classification is far from complete; Ahmad and Saksena (1981) have measured isotope shifts in over one hundred spectral lines as an aid to this classification. This is the latest example of the way in which isotope shift data have played an important role in the interpretation of Nd I and Nd II.

9.61. PROMETHIUM

Promethium has no stable isotope, and there is nothing to report.

9.62. SAMARIUM

Samarium has been mentioned in earlier chapters of this book. It has had a long and distinguished career in the ranks of elements of interest to workers in the field of isotope shifts. Interest has not slackened in recent years; if anything it has intensified. Isotope shifts in electronic x-ray spectra have been measured by Lee and Boehm (1973). Muonic x-ray energies have been measured and interpreted in terms of nuclear charge distributions by Yamazaki *et al.* (1978), Powers *et al.* (1979), and Barreau *et al.* (1981), who among them have studied isotopes from ^{144}Sm to ^{154}Sm, inclusive. Moinester *et al.* (1981) have made a combined analysis using elastic electron scattering data and muonic data. In the field of optical isotope shifts, a *J*-dependent isotope shift has been measured (Bauche *et al.*, 1977) in the ground Russell–Saunders term of Sm I, which has the same sign and order of magnitude as an *ab initio* evaluated value of the field shift when the crossed second-order effect of the atomic magnetic interactions is included. Laser atomic beam spectroscopy has been used by Brand *et al.* (1980) to measure shifts in many lines from which they deduce values of $\delta \langle r^2 \rangle$, the change between isotopes in the mean square radius of the nuclear charge distribution. Their work was of sufficient precision to show the hyperfine perturbation of the isotope shifts of the odd isotopes that had been predicted beforehand by Labarthe (1978). These perturbations reveal themselves because the points involving odd isotopes in a King plot do not lie on the straight line on which the shifts between even isotopes lie. The deviations in the shifts in $\lambda 590.26$ nm were up to about 0.5 MHz in measured shifts of a few hundred MHz. These shifts were measured with an uncertainty of about 0.3 MHz. Griffith *et al.* (1979) had measured shifts to similar accuracy using optical heterodyne methods, but had looked at the even isotopes only. These gave some anomalous shifts, i.e., ones which did not lie on a straight line in a King plot, which arose from mixing between closely spaced levels (Griffith *et al.*, 1981). A theoretical treatment of this effect based on diagonalization of the Hamiltonian matrix describing the two mixed levels was given by Palmer and Stacey (1982).

The changes in the mean square radius of the nuclear charge distribution have been determined by Brand *et al.* (1980) from their optical isotope shift measurements in numerous $6s^2$–$6s6p$ transitions. All the evidence from spectra in this and neighboring elements points to a small specific mass shift in such transitions. A Hartree–Fock evaluation (Bauche, 1974) gives a Vinti k factor of -0.03 a.u., i.e., a specific mass shift of -9 MHz between ^{154}Sm and ^{152}Sm. Brand *et al.* (1980) assumed the specific mass shift to be 0(12) MHz in $\lambda 591.64$ nm, and from this determined the $\delta \langle r^2 \rangle$ values given in Table 9.15. Since the field shifts are about 1000 MHz, this assumption introduces little error into the $\delta \langle r^2 \rangle$ values. Radii deduced from muonic x-ray studies are also given in Table 9.15.

Optical isotope shifts have been measured to a similar level of precision by Griffith *et al.* (1979) who used an heterodyne method. Their most interesting discovery was that the shifts between even isotopes in two spectral lines were anomalous in that they did not lie on a straight line in a King plot when plotted against the shifts in any other spectral line. Griffith *et al.* (1981) showed that the anomalies arise from two closely spaced levels that are interacting with each other. The two levels are $19,175$ cm^{-1} and $19,192$ cm^{-1} above the ground state, respectively. Transitions to these two levels

TABLE 9.15
Charge Radii of Samarium Nuclei

Mass number	$\langle r^2 \rangle$(fm^2)[a]	$\delta \langle r^2 \rangle$(fm^2)	
		from previous column	from optical isotope shifts[b]
144	24.52(6)		
		0.49(9)	0.517(27)
148	25.01(7)		
		0.34(11)	0.303(16)
150	25.35(8)		
		0.49(11)	0.423(22)
152	25.84(7)		
		0.31(13)	0.230(12)
154	26.14(11)		
147	24.87(6)		
		0.14(9)	0.152(8)
148	25.01(7)		
		0.08(10)	0.092(5)
149	25.09(7)		

[a] From muonic x-ray studies of Powers *et al.* (1979) and Barreau *et al.* (1981).
[b] From Brand *et al.* (1980).

were found to have equal but opposite anomalous shifts. Transitions not involving these levels showed no anomalous shifts. The anomalous shifts can be displayed by plotting modified shifts in the fictitious transition between the two closely spaced interacting levels against modified shifts in a transition involving neither of these levels. Results for this fictitious transition are available because transitions from both closely spaced levels to a common lower level were studied. Such a King plot shows deviations from a straight line of up to at least 30 MHz, but the question is, from which straight line are the shifts deviating? Griffith *et al.* (1981) present their results with the shift for ^{154,152}Sm deviating considerably from a line which passes much closer to the shifts for the other pairs of isotopes. King (1981) suggested that a line could be found so that the deviations varied smoothly with mass but not smoothly with field shift and deduced from this that the anomalous shifts are mass shifts. Palmer and Stacey (1982) arbitrarily chose one isotope, and treated the two closely spaced levels for this isotope as an unperturbed system. The Hamiltonian matrix is

$$\begin{pmatrix} E & 0 \\ 0 & 0 \end{pmatrix}$$

where E is the separation between the levels. Using the electronic wavefunctions of the chosen isotope, A_0, as a basis, the Hamiltonian matrix for another isotope, A, is

$$\begin{pmatrix} E + f_1 \delta \langle r^2 \rangle_{AA_0} + m_1 \alpha_{AA_0} & f_{12} \delta \langle r^2 \rangle_{AA_0} + m_{12} \alpha_{AA_0} \\ f_{12} \delta \langle r^2 \rangle_{AA_0} + m_{12} \alpha_{AA_0} & f_2 \delta \langle r^2 \rangle_{AA_0} + m_2 \alpha_{AA_0} \end{pmatrix}$$

where

$$\alpha_{AA_0} = \frac{M_A - M_{A_0}}{M_A M_{A_0}} \tag{9.4}$$

If the off-diagonal elements were zero the isotope shift would be

$$\beta_{AA_0} = (f_1 - f_2) \delta \langle r^2 \rangle_{AA_0} + (m_1 - m_2) \alpha_{AA_0} \tag{9.5}$$

and with the inclusion of off-diagonal elements the isotope shift is

$$\left[(E + \beta_{AA_0})^2 + 4 \left(f_{12} \delta \langle r^2 \rangle_{AA_0} + m_{12} \alpha_{AA_0} \right)^2 \right]^{1/2} - E$$

Since shifts have been measured for four pairs of isotopes, the four unknowns

$f_1 - f_2$, $m_1 - m_2$, f_{12}, and m_{12} can be determined. In the absence of off-diagonal elements the isotope shift of Eq. (9.5) is a sum of mass and field shifts, but when off-diagonal elements are included, such a division is impossible. There is thus little point in discussing whether the anomalies are mass effects or field effects. Furthermore, on this approach, there is no one unique straight line on the King plot from which the shifts are deviating; the shifts without mixing, β_{AA_0}, depend on which isotope, A_0, is chosen to be the unperturbed system.

The J-dependence of isotope shifts (Bauche *et al.*, 1977) has been measured very precisely by New *et al.* (1981). They studied transitions to seven levels of the ground term of Sm I, $4f^6 6s^2\,{}^7F$. They were able to obtain isotope shifts for fictitious transitions between the levels of this term for four different pairs of even isotopes. A plot of these shifts against those in an actual spectral line allowed the shifts between J values to be divided into mass shifts and field shifts, and showed that the J-dependent field shifts were at least four times larger than the J-dependent mass shifts.

9.63. EUROPIUM

The relatively simple spectrum of europium stands out as an oasis to the atomic spectroscopist in the desert of complicated spectra of most other rare earths. Eu I and II are well analyzed and, with pure configurations available, it has been possible to deduce a value for the change between ^{153}Eu and ^{151}Eu in the mean square radius of the nuclear charge distribution from optical isotope shifts with considerable confidence. Brand *et al.* (1981) studied $4f^7 5d6s$–$4f^7 5d6p$ transitions in Eu I. An atomic beam of europium was crossed at right angles by a beam from a tunable dye laser. Interaction was detected by the resonance fluorescence perpendicular to both beams, and the $4f^7 5d6s$ level was populated by running an arc discharge in the atomic beam. The electron density at the nucleus

$$\pi |\psi(0)|^2 (6s \text{ in } 4f^7 5d6s) a_0^3 / Z = 0.258 \qquad (9.6)$$

was obtained from magnetic hyperfine structure. The screening factor

$$\beta = \frac{\Delta|\psi(0)|^2 (4f^7 5d6s\text{–}4f^7 5d6p)}{|\psi(0)|^2 (6s \text{ in } 4f^7 5d6s)} = 1.14 \qquad (9.7)$$

was taken from a relativistic Dirac–Fock evaluation of Coulthard (1973). The specific mass shift in the transition is very small compared with the field shift and so an accurate value of $\delta\langle r^2 \rangle$ can be obtained. By combining their results with those obtained from other transitions (Zaal *et al.*, 1979) Brand *et al.*, (1981) deduce that for ^{153,151}Eu, $\delta\langle r^2 \rangle = 0.577(25)$ fm^2. Wilson (1982) has made a pseudorelativistic Hartree–Fock evaluation of the specific mass shift, obtaining the value -24 MHz, which is indeed very small compared with the field shift of $-3546(3)$ MHz. Kronfeldt *et al.* (1982) considered the ratios of field shifts in various transitions and pointed out that these are in better agreement with the relativistic Dirac–Fock evaluation of the ratios of $\Delta|\psi(0)|^2$ of Coulthard (1973) than with the nonrelativistic Hartree–Fock evaluation of Wilson (1972). Wilson (1982) showed that the pseudorelativistic Hartree–Fock evaluation of these ratios (see Section 4.2.2) also compares well with the experimental values and provides a useful and simpler alternative to the relativistic Dirac–Fock approach.

A small J dependence of the isotope shift, a crossed second-order effect, has been observed in the 4f^{7}5d6s a^{10}D term by Kronfeldt *et al.* (1982) using interferometric investigations, and in this term and the a^8D term by Pfeufer *et al.* (1982) using laser–atomic-beam spectroscopy.

Isotope enrichment by a laser process which uses charge transfer ionization has been demonstrated in europium and hydrogen by Boerner *et al.* (1978).

9.64. GADOLINIUM

There is little recent work to report; the $\delta\langle r^2\rangle$ values of Heilig and Steudel (1974) have not been superseded. $\delta\langle r^2\rangle/\text{fm}^2$: ^{160,158}Gd, 0.140(15); ^{158,156}Gd, 0.135(20); ^{156,154}Gd, 0.174(24); ^{154,152}Gd, 0.407(45); ^{157,156}Gd, 0.021(13); ^{157,155}Gd, 0.106(28); ^{155,154}Gd, 0.089(20). Odintsova and Striganov (1976) measured isotope shifts in 65 optical transitions. They estimated the nuclear deformation of the rare isotope ^{152}Gd from their relative isotope shifts. Ahmad *et al.* (1979) have measured the ^{160,156}Gd isotope shift in hundreds of lines of Gd I and Gd II in the wavelength region between 414 nm and 454 nm. They deduced the electronic configurations of many previously unassigned levels on the basis of their observed isotope shifts.

9.65. TERBIUM

Terbium has only one stable isotope, and there is nothing to report.

9.66. DYSPROSIUM

The very complex arc and spark spectra of dysprosium have been analyzed by Wyart, and his analysis has been used by Zaal *et al.* (1980) to determine the isotope shifts of pure configurations from the isotope shifts that they measured in 31 transitions of Dy I. They deduced the specific mass shifts

TABLE 9.16

Isotope Shifts for ^{164,162}Dy in Dy I[a]

Transition between the pure configurations	Specific mass shift (MHz)	Field shift (MHz)
$4f^{10}6s^2 - 4f^{10}6s6p$	7(8)	−992(8)
$4f^{10}6s^2 - 4f^96s^25d$	−529(14)	1646(14)
$4f^{10}6s^2 - 4f^96s5d^2$	−475(25)	230(25)

[a] Figures obtained from Zaal *et al.* (1980).

and field shifts given in Table 9.16. They also deduced the changes between isotopes in the mean square radius of the nuclear charge distribution given in Table 9.17. Attempts to disentangle the configuration mixing in Dy II by a study of isotope shifts have proved less satisfactory (Aufmuth, 1978). Aufmuth showed that crossed second-order effects and relativistic effects are present. The former were revealed by a difference between the isotope shifts of the ^{4}I and ^{6}I terms, and the latter by a *J*-dependent isotope shift in the ^{6}I term of the $4f^{10}6s$ configuration. Although these spectra are still not completely analyzed, the study of isotope shifts has made significant contributions to our understanding of them.

Isotope separation by a laser process which uses charge transfer ionization by Cs$^+$ ions has been demonstrated by Burghardt *et al.* (1980).

9.67. HOLMIUM

Holmium has only one stable isotope, and there is nothing to report.

9.68. ERBIUM

The isotope shifts of many spectral lines have been measured and used as an aid to the analysis of Er I (Miller and Ross, 1976). There is considerable

TABLE 9.17

Mean Square Radii of the Charge Distribution of Dysprosium Nuclei Relative to That of ^{162}Dy(fm^2)[a]

Mass number	164	163	161	160	158	156
$\delta \langle r^2 \rangle$	+0.121 (10)	+0.040 (3)	−0.097 (8)	−0.132 (10)	−0.262 (15)	−0.453 (20)

[a] Figures from Zaal *et al.* (1980).

configuration mixing so, although the measured shifts are consistent with results obtained in the spectra of neighboring elements, definite quantitative conclusions cannot be drawn. Heilig and Steudel (1974) gave lower limits for values of the changes between isotopes in the mean square radius of the nuclear charge distribution.

9.69. THULIUM

Thulium has only one stable isotope, and there is nothing to report.

9.70. YTTERBIUM

Clark *et al.* (1979) have measured isotope shifts in $\lambda 555.6$ nm with great precision using a cw tunable dye laser and atomic beam techniques. By comparing their results with shifts in electronic and muonic x-ray spectra they conclude that the specific mass shift for ^{176,174}Yb is about 250 MHz in $\lambda 555.6$ nm. This corresponds to a Vinti k factor of 1 a.u. Such large specific mass shifts are produced by a change in the number of 4f electrons and a positive shift means less 4f electrons in the lower level or more in the upper level. The transition is nominally $4f^{14}6s^2\,^1S_0 - 4f^{14}6s6p\,^3P_1$, so the large specific mass shift must arise from a mixture of a $4f^{13}$ configuration into the lower level. There is no other evidence for this. All the spectroscopic evidence is that the lower level, which is the ground level, is from a pure $4f^{14}6s^2$ configuration. This suggests that Clark *et al.* (1979) have given too much weight to the muonic x-ray results, a suggestion which is confirmed when the optical isotope shift work of Grundevik *et al.* (1979) is taken into account. They measured shifts in $\lambda 398.8$ nm $4f^{14}6s^2\,^1S_0 - 4f^{14}6s6p\,^1P_1$ and a comparison of the two sets of results shows that

$$FS_{398.8} = 0.456(3)\, FS_{555.6} \qquad (9.8)$$

Such a small ratio for the field shifts indicates that the nominal configurations cannot all be pure. The spectroscopic evidence (Martin *et al.*, 1978) is that it is the $^1P_1$ level which is impure. A mixture of $4f^{13}5d6s^2$ would reduce the field shift in $\lambda 398.8$ nm as required, and so is a likely possibility. It would also introduce a negative specific mass shift into $\lambda 398.8$ nm since the transition would involve less 4f electrons in the upper level. The relation between the specific mass shifts in MHz for ^{176,174}Yb is

$$SMS_{398.8} = 0.456(3)\, SMS_{555.6} - 91(3) \qquad (9.9)$$

So a specific mass shift of about -90 MHz in $\lambda 398.8$ nm, which is compatible with a large admixture of $4f^{13}5d6s^2$ in $4f^{14}6s6p\,^1P_1$, is consistent

TABLE 9.18

Isotope Shifts and Changes in Nuclear Charge Radius for Ytterbium Isotopes

Mass numbers of isotope pair	Field shift plus specific mass shift (MHz)[a]	$\delta\langle r^2\rangle$(fm^2)			
		b	c	d	e
176,174	−974.00(50)	0.089	0.109(8)	0.107(5)	0.104(11)
174,172	−1019.96(50)	0.034	0.114(8)	0.115(5)	0.138(12)
172,170	−1306.60(50)	0.120	0.139(8)	0.145(4)	0.167(16)
170,168	−1389.23(50)	0.127	0.147(8)		
174,173	−565.39(50)	0.052	0.061(4)	0.065(7)	0.088(27)
172,171	−835.77(50)	0.077	0.085(4)	0.083(6)	0.090(32)
169,168		0.056 [f]			

[a] Measurements of Clark et al. (1979) in λ555.6 nm, $4f^{14}6s^2\,^1S_0$–$4f^{14}6s6p\,^3P_1$.

[b] Assuming the transition is between pure configurations.

[c] Values deduced by Clark et al. (1979).

[d] From the values of $\delta\langle r^2\rangle$ quoted by Barrett and Jackson (1977) from the measurements of Zehnder et al. (1975) of muonic x-ray energies.

[e] From electronic x-ray isotope shifts (Lee and Boehm, 1973).

[f] From the work of Champeau et al. (1974) who obtained ^{169}Yb by a (n, γ) reaction from ^{168}Yb and observed the spectrum from a hollow cathode lamp.

with zero specific mass shift in λ555.6 nm, which is compatible with this transition being between pure configurations. The change between isotopes in the mean square radius of the nuclear charge distribution can thus be determined from isotope shifts in λ555.6 nm alone. Using (a) the very accurate results of Clark et al. (1979), (b) their estimate from magnetic hyperfine splittings of the conversion factor between field shift and $\delta\langle r^2\rangle$ of 10,900 MHz/fm^2 and (c) a specific mass shift of zero (based on the above arguments), the values of $\delta\langle r^2\rangle$ in the third column of Table 9.18 are obtained. The values of $\delta\langle r^2\rangle$ obtained by Clark et al. (1979), who gave more weight to the muonic energies and took the specific mass shift to be about 250 MHz, are given in the fourth column. The results which can be deduced from muonic x-ray energy measurements (Zehnder et al., 1975) are given in the fifth column. There is clearly a discrepancy between the optical and the muonic results if it is accepted that λ555.6 nm is a transition between pure configurations. Isotope shift measurements in electronic x-ray spectra (Lee and Boehm, 1973) lead to values of $\delta\langle r^2\rangle$ which are much closer to the muonic than to the optical results; they are given in the sixth column of Table 9.18.

9.71. LUTETIUM

The isotope shift between the two naturally occurring isotopes has been measured by Zimmermann et al. (1980) in two optical transitions with the aid of a tunable dye laser and an atomic beam of lutetium. The spectrum is complex and there is much configuration mixing, but with the aid of the work of Wyart a value for the change in the mean square radius of the nuclear charge distribution for ^{176,175}Lu has been deduced; $\delta\langle r^2 \rangle = 0.022(5)$ fm^2, a remarkably small value.

9.72. HAFNIUM

Optical isotope shifts have been measured in several transitions by Čajko (1970) who deduced that the isotope shift for ^{180,178}Hf in a pure $5d^2 6s^2 - 5d^2 6s6p$ transition is 1353(9) MHz and that the shift due to a 6s electron is 2000(350) MHz. From this and other work Heilig and Steudel (1974) deduced the following values of the change in nuclear radius, $\delta\langle r^2 \rangle / $fm^2: ^{180,178}Hf, 0.078(16); ^{178,176}Hf, 0.065(14); ^{176,174}Hf, 0.068(14); ^{179,178}Hf, 0.031(13); ^{177,176}Hf, 0.008(8).

9.73. TANTALUM

Some isotope shift work was done at Bonn University (see Heilig, 1982), but has not yet been published in an accessible journal.

9.74. TUNGSTEN

There has been no recent work to supersede the figures for nuclear charge radii given by Barrett and Jackson (1977). These are reproduced in Table 9.19.

Champeau and Miladi (1974) measured isotope shifts in two optical transitions, $\lambda426.9$ nm ($5d^5 6s - 5d^3 6s^2 6p$) and $\lambda498.2$ nm ($5d^4 6s^2 - 5d^4 6s6p$).

TABLE 9.19
Changes in the Mean Square Radius of the Charge Distribution for Tungsten Nuclei

Mass numbers of isotope pair	$\delta\langle r^2 \rangle$(fm^2)		
	a	b	c
186,184	0.080(14)	0.092(5)	0.066(9)
184,182	0.091(16)	0.120(5)	0.102(12)
182,180	0.064(12)		
183,182	0.048(10)		

[a] Results from optical isotope shifts of $\beta\delta\langle r^2 \rangle$ where β is the screening factor of Eq. (4.33) (Heilig and Steudel, 1974).
[b] Results from muonic x-ray energies (Hitlin et al., 1970).
[c] Results from isotope shifts in x-ray spectra of electronic atoms (Chesler and Boehm, 1968).

They obtained the following relation between the specific mass shifts in MHz for ^{186,184}W

$$SMS_{498.2} + 0.61(2)SMS_{426.9} = -120(60) \tag{9.10}$$

Assuming the specific mass shift is entirely due to changes in the number of 5d electrons, then the specific mass shifts in the two lines are -200 MHz and zero. The large specific mass shift in $\lambda426.9$ nm corresponds to a Vinti k factor of -0.95 a.u. A Hartree–Fock evaluation gives a k factor of about -1.3 a.u. for a change of two in the number of 5d electrons. Because of the large extrapolation error in obtaining the experimental result, the two results are not inconsistent.

9.75. RHENIUM

The first spectrum of rhenium is a good example of the use of isotope shifts to classify a complex spectrum of a heavy element. Rhenium is sufficiently heavy for field shifts to be much larger than mass shifts and so the latter can be ignored for purposes of spectral classification. Buchholz *et al.* (1979) measured isotope shifts in optical transitions between the ground level $5d^5 6s^2\,^6S_{5/2}$ and excited levels of what was thought to be the $5d^4 6s^2 6p$ configuration. With $6s^2$ in both upper and lower levels, the field shift arises from the difference in the screening of 5d compared with 6p. The screening of 5d is larger, so the field shift is larger in the upper level. It follows that the field shift in the transition should be positive. This was true for only a minority of the lines studied. Buchholz *et al.* (1979) were able to deduce the large amounts of $5d^5 6s6p$ (and possibly $5d^6 6p$) which are present in the upper levels of the lines with negative field shifts. Bürger *et al.* (1982) have made a parametric study of the isotope shifts that they measured in several optical transitions in Re I, and were able to show the presence of second-order (J-dependent) field shifts with parameter values in good agreement with those obtained for other 5d-shell atoms (see, for example, Grethen *et al.*, 1980).

9.76. OSMIUM

As in rhenium, sizes of isotope shifts have been used to classify the first spectrum (Gluck *et al.*, 1964). Such work has been combined with measurements of relative isotope shifts between various pairs of isotopes (Hines and Ross, 1962; Nöldeke *et al.*, 1962) to obtain values of $\delta\langle r^2 \rangle$ (Heilig and Steudel, 1974). These values do not agree with those obtained from a study of muonic x-ray energies by Hoehn *et al.* (1981). The discrepancy probably

TABLE 9.20

Charge Radii of Osmium Nuclei

Mass number	$\langle r^2 \rangle$(fm^2)[a]	$\delta \langle r^2 \rangle$(fm^2)[a] from previous column	Relative field shifts[b]	$\delta \langle r^2 \rangle$(fm^2) from comparison of previous two columns
184				
			1.31(4)	0.107(4)
186	28.927			
		0.104(4)	1.29(2)	
188	29.031			
		0.090(4)	1.12(2)	
190	29.121			
		0.085(1)	1	
192	29.206			
186				
			0.38(4)	0.031(4)
187				
188				
			0.36(4)	0.029(4)
189				

[a] From muonic x-ray studies of Hoehn *et al.* (1981).
[b] Figures from Heilig and Steudel (1974).

arises from an inaccurate conversion from optical field shift to the change in the mean square radius of the nuclear charge distribution. The relative isotope shifts are known fairly accurately, and so, since the mass shift components of these are small, they can be treated as relative field shifts. These relative field shifts can be used to extend the values of $\delta \langle r^2 \rangle$ obtained from muonic x-ray energy measurements to the other isotopes for which optical isotope shifts have been measured. The results are given in Table 9.20.

9.77. IRIDIUM

No recent work has yet been published in an accessible journal; see Heilig and Steudel (1974) and Heilig (1982).

9.78. PLATINUM

Baird and Stacey (1974) have measured optical isotope shifts in three lines. They found no evidence for large specific mass shifts, and, assuming them to be negligible, deduced the following relative field shifts from the shifts in the line they measured most accurately, $\lambda 444.2$ nm $5d^8 6s^2\,^3F_3 - 5d^9 6p\,^3P_2$:

$192,190$ $0.905(28)$; $194,192$ $0.969(7)$; $196,194$ 1; $198,196$ $1.084(4)$; $195,194$ $0.456(4)$.

Müller and Winkler (1975) deduced from their study of $5d^8 6s^2$ and $5d^9 6s$ that βC_{exp} for ^{196,194}Pt is $158(15)$ mK. Using Eq. (4.25) and Table 4.2 it follows that $\beta \delta \langle r \rangle^2$ is 0.082 fm^2 for ^{196,194}Pt. Other values of the change in mean square radius of the nuclear charge distribution can be obtained directly from the relative field shifts given above. The error in ignoring the specific mass shift is small compared with the uncertainty in β. The work of Rajnak and Fred (1977) on actinide configurations shows that any successful evaluation of β for an element as heavy as platinum will have to include relativistic effects. There is also the problem of ensuring that the isotope shifts are measured between pure configurations, or at least between levels for which the amount of configuration mixing is known. Grethen et al. (1980) have used isotope shift results to make a parametric study of the low even levels of Pt I and have obtained estimates of the amount of mixing of $5d^9 6s$ and $5d^8 6s^2$ in various levels. No similar work has yet been carried out on the odd levels.

9.79. GOLD

Gold has only one stable isotope, ^{197}Au, and no isotope shift work had been carried out until very recently. Kluge et al. (1983) have measured the shift in the $\lambda 267.6$ nm $6s\,^2S_{1/2} - 6p\,^2P_{1/2}$ line between ^{197}Au and ^{195}Au, which has a half-life of 183 days. (AUTHOR'S NOTE: I will not comment on this work, since it is merely preliminary to the study of many shorter lived isotopes studied on-line at the ISOLDE facility at CERN. The intention is to build up a comparable set of data to that which is already available in the case of mercury, the next element in the Periodic Table.)

9.80. MERCURY

The first spectrum of mercury is relatively simple, but one of the earliest studies of isotope shifts (Schüler and Jones, 1932) showed that the behavior of some levels of the odd isotopes is not entirely straightforward. The levels $6s6d\,^1D_2$ and $6s6d\,^3D_1$ are only 3 cm^{-1} apart and, since the hyperfine splittings of the odd isotopes are of a comparable size, there is a hyperfine perturbation of one fine structure level by another. This effect is also mentioned in Section 6.1 and Section 9.3.1; it is often referred to as the second-order hyperfine interaction. The interaction of the two levels has been carefully analyzed by Gerstenkorn and Vergès (1975). They studied infrared transitions by Fourier transform spectroscopy and found that the relative isotope shifts were anomalous in the transitions involving the perturbing levels mentioned above. In many transitions with large field shifts the isotope shift between ^{200}Hg and the center of gravity of ^{199}Hg is 0.786 of the ^{202,200}Hg shift. This can be taken as the ratio of the field shifts (since mass shifts are very much smaller than field shifts in these transitions) for unperturbed levels. In the line $\lambda 1.814$ μm ($5d^{10}6s6d\,^3D_3 - 5d^9 6s^2 6p^3 F_4$) this ratio was found to be 0.811 rather than 0.786. The actual shift between ^{200}Hg and the center of gravity of ^{199}Hg is 5190 MHz and so the second-order shift of the center of gravity is 160 MHz. As the nuclear spin of ^{199}Hg is $\frac{1}{2}$, there are two hyperfine components of $^3D_3$ with F $= \frac{5}{2}$ and $\frac{7}{2}$. There is no hyperfine level of $^1D_2$ with F $= \frac{7}{2}$, and so all the hyperfine perturbation is of the $^3D_3$ F $= \frac{5}{2}$ level. Since the center of gravity is shifted by 160 MHz, it follows that the F $= \frac{5}{2}$ level is shifted by 380 MHz. This second-order hyperfine interaction is also responsible for the $\lambda 296.76$ nm ($6s6p^3P_0 - 6s6d\,^1D_2$) line not being completely forbidden. Coillaud *et al.* (1980) observed this line and, as would be expected, it appeared only for the odd isotopes having one component for ^{199}Hg and three for ^{201}Hg (which has a nuclear spin of $\frac{3}{2}$). Gerstenkorn *et al.* (1977) have made further studies of this interaction and show that it can be explained by empirically introducing a hyperfine parameter into the off-diagonal matrix element between $^1D_2$ and $^3D_2$; there is no need to introduce any configuration mixing.

Even in transitions in which there is no second-order hyperfine interaction, the behavior of the odd isotopes is still of considerable interest. The earliest measurements had shown that, as in several other elements, the isotopes exhibited odd–even staggering in their relative positions in the spectrum; the center of gravity of the spectral components of an isotope with an odd number of neutrons (odd N) was found to lie closer to the line of the adjacent lighter even-N isotope than to the line of the adjacent heavier even-N isotope. It is a field shift effect and is now interpreted as arising because the change in the mean square radius of the nuclear charge distribution is smaller when a neutron is added to an even-N isotope than to an odd-N isotope. The effect is usually characterized by the staggering parameter

$$\gamma_{(N+1)} = \frac{2\left(\langle r^2 \rangle_{N+1} - \langle r^2 \rangle_N\right)}{\langle r^2 \rangle_{N+2} - \langle r^2 \rangle_N} \tag{9.11}$$

where N (the number of neutrons) is even. The staggering parameter would be equal to unity in the absence of staggering but is often less than unity and can even be negative.

The first measurement of an isomer shift in any element was made in $\lambda 253.7$ nm ($6s^2\,^1S_0 - 6s6p\,^3P_1$) of Hg I by Melissinos and Davis (1959). They observed the spectrum of ^{197}Hg in both its ground state with $I = \frac{1}{2}$ and a half-life of 65 hr, and in an excited isomeric state with $I = \frac{13}{2}$ and a half-life of 25 hr. They obtained the following staggering parameters: $^{197}\gamma = 0.67$ and $^{197m}\gamma = 0.98$. The lack of staggering for isomeric ^{197m}Hg with $I = \frac{13}{2}$ was explained by Tomlinson and Stroke (1962) who measured isotope shifts in other radioactive isotopes and isomers of mercury. They pointed out that the neutrons in the even-N isotopes of mercury are in $i_{13/2}$ orbits, so the lack of odd–even staggering for the isomeric states, which also have the last (odd-N) neutron in an $i_{13/2}$ orbit, is because an $i_{13/2}$ neutron has the same effect on the nuclear charge radius whether it is a paired or an unpaired neutron. The odd–even staggering arises because the odd neutrons in the ground states of odd isotopes are in lower angular momentum orbits. More recent work (Stroke *et al.*, 1979) has shown trends in the staggering for isomers from ^{185m}Hg to ^{199m}Hg, which is discussed in Section 10.2.

Much recent work on optical isotope shifts in mercury has concentrated on measurements of the spectra of radioactive neutron deficient isotopes. Bonn *et al.* (1976) measured isotope shifts between odd isotopes down to ^{181}Hg. The general trend within the stable isotopes continues to ^{187}Hg, but between ^{187}Hg and ^{185}Hg there is a sharp discontinuity which is thought to be a deformation effect. The neutron deficient isotopes were produced at the on-line mass separator ISOLDE at CERN. The radioactive isotope was transferred to an optical pumping apparatus. The orientation built-up by optical pumping was monitored via the asymmetry or anisotropy of the β rays emitted by the radioactive isotopes. Zeeman scanning of a mercury isotope lamp was used to reach resonance with the hyperfine components of the $\lambda 253.7$ nm line of the investigated isotope. This nuclear radiation detected optical pumping (RADOP) method, which is described by Huber *et al.* (1976), can only be used for isotopes with a finite nuclear spin. The development of a purely optical technique which could be used for even isotopes with zero nuclear spin showed that there is not only a discontinuity in the isotope shifts at about ^{186}Hg but that there are enormous amounts of odd–even staggering present (see Fig. 9.10). Whereas previously the staggering parameter had varied from one to about zero, Kühl *et al.* (1977) obtained $^{185}\gamma = -11$ from their measurements on ^{184}Hg and ^{186}Hg. The wavelengths were measured by resonance fluorescence of the $\lambda 253.7$ nm line with laser UV radiation produced by frequency doubling in a crystal of the output of a tunable dye laser. This technique has also been used to measure isomer shifts (Dabkiewicz *et al.*, 1979). The interpretation of the results in terms of nuclear deformations is discussed in Section 10.2.

Before this can be done, changes in the nuclear charge distribution must be determined from the optical isotope shifts. This determination is sufficiently precise, and mercury is a sufficiently heavy element, that the theory of Seltzer (1969), described in Section 5.1, should be used. He showed that field shifts are not proportional to $\delta\langle r^2 \rangle$ exactly but to λ as given by Eq. (5.4).

The calculation of λ from isotope shifts in $\lambda 253.7$ nm ($6s^2\,^1S_0 - 6s6p\,^3P_1$) is straightforward; there is no second-order hyperfine interaction and the specific mass shift is very small. Bauche (1974) evaluated k to be 0.08 a.u. for a similar transition in osmium. Such a value of k gives a specific mass shift of 14 MHz for ^{202,200}Hg. As the isotope shifts are about 2000 MHz, little error is introduced by assuming the specific mass shift to be 0(14) MHz. The electron densities at the nucleus can be written

$$\Delta|\psi(0)|^2(6s^2 - 6s6p) = \beta|\psi(0)|^2\big(6s \text{ in }^2S_{1/2} \text{ of Hg II}\big) \qquad (9.12)$$

where β has been evaluated by Wilson (1968) to be 0.74. [See Section 4.2.2, particularly Table 4.5 and Eq. (4.35).] Wilson used Hartree–Fock wavefunctions, but relativistic effects should be taken into account for an element as heavy as mercury. Such relativistic evaluations have not been made for mercury, but for the similar factor in europium, $(4f^76s^2 - 4f^76s6p)/(6s$ in $4f^76s)$, a relativistic evaluation (Coulthard, 1973) gave 0.73, a very similar value to the Hartree–Fock one of 0.71 (Wilson, 1972).

Let us assume that, for mercury, the relevant β is 0.74(5). Then $\pi a_0^3|\psi(0)|^2$ (6s in $^2S_{1/2}$) can be determined from either the Goudsmit–Fermi–Segrè formula or magnetic hyperfine splitting. The former gives 80.7, the latter 82.3; let us assume 82(2). The $f(z)$ of Eq. (4.24) is, from Table 4.2, 66.1 GHz fm^{-2}. The isotope shift in $\lambda 253.7$ nm for ^{202,200}Hg is $-5300(20)$ MHz, so the field shift is $-5330(24)$ MHz and λ is 0.106(8) fm^2. [Seltzer's theory is incorporated into Eq. (4.24) by replacing $\delta\langle r^2\rangle$ by λ.] Bonn et al. (1976) obtained a rather larger value for λ of 0.118 fm^2 because they took β (they called it γ) to be 0.67.

The uncertainty in λ arises predominantly from the uncertainty in our knowledge of the screening factor β. This uncertainty does not arise in electronic x-ray isotope shifts, and so it is instructive to compare the x-ray shift measurements of Lee et al. (1978) for five mercury isotopes with the optical results. A plot of their λ values against the optical field shifts with $\lambda = 0$ when $FS = 0$ is shown in Fig. 9.8. Using the more accurate relative optical shifts to smooth the x-ray values of λ gives λ for ^{202,200}Hg as 0.099(6) fm^2. This independent assessment suggests the screening factor β should be larger than the 0.74 used above. Let us combine the two results for λ as

$$\lambda \text{ for } ^{202,200}\text{Hg} = 0.102(6) \text{ fm}^2 \qquad (9.13)$$

Thompson et al. (1983) have estimated that for lead

$$\lambda = 0.93(2)\delta\langle r\rangle^2 \qquad (9.14)$$

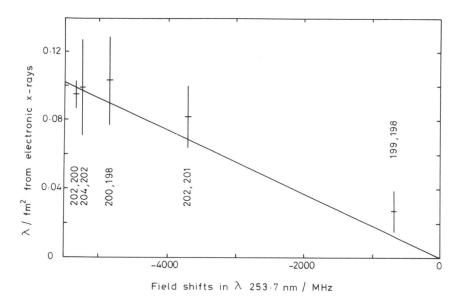

FIGURE 9.8. λ from electronic x-ray isotope shifts plotted against optical field shifts in mercury. The λ values are those of Lee *et al.* (1978). The field shifts have been obtained from the isotope shifts given by Bonn *et al.* (1976) with the assumption that the specific mass shift is zero.

so assuming the same relationship for mercury

$$\delta\langle r^2\rangle \text{ for } {}^{202,200}\text{Hg} = 0.110(6) \text{ fm}^2 \qquad (9.15)$$

Another independent assessment of $\delta\langle r^2\rangle$ can be obtained from the muonic x-ray transition study of Hahn *et al.* (1979). From the 1s–2p transitions they determined the model-independent, equivalent uniform radius R_k of Eq. (5.5), with $k = 2.413$, and its difference between isotopes, δR_k. They also gave the more model-dependent c parameters of a Fermi distribution (see Fig. 4.1) with $a = 0.524$ fm for all isotopes. From these, values of $\delta\langle r^2\rangle$ can be obtained which are plotted against the optical field shifts in Fig. 9.9 with $\delta\langle r^2\rangle = 0$ when $FS = 0$. The shift involving ^{201}Hg lies well off the overall trend, probably because of difficulty in determining the nuclear polarization of this isotope by the muon; the difficulty arises from the large number of nuclear levels near the ground state for this isotope. Ignoring this point, the best-fit line shown in Fig. 9.9 gives

$$\delta\langle r^2\rangle \text{ for } {}^{202,200}\text{Hg} = 0.106(3) \text{ fm}^2 \qquad (9.16)$$

The uncertainty of 3 is deduced from the scatter of the results and something must be added for the model-dependent nature of the analysis. Treating the

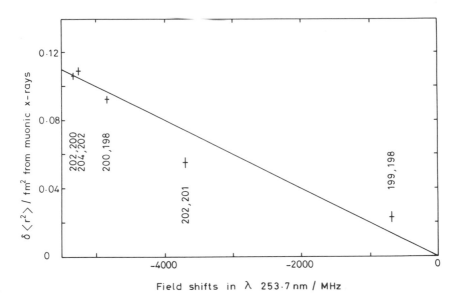

Field shifts in λ 253·7 nm / MHz

FIGURE 9.9. $\delta\langle r^2 \rangle$ from muonic x-ray energies plotted against optical field shifts in mercury. The $\delta\langle r^2 \rangle$ values have been obtained from the Fermi distribution parameters of Hahn *et al.* (1979). The uncertainties are based on their uncertainties in the model-independent quantity R_k and so underestimate the true uncertainties. The field shifts are as in Fig. 9.8.

results of Eqs. (9.15) and (9.16) with equal weight gives the final result

$$\delta\langle r^2 \rangle \text{ for } {}^{202,200}\text{Hg} = 0.108(5) \text{ fm}^2 \qquad (9.17)$$

It is perhaps worth stressing that the three independent results are mutually consistent: $\delta\langle r^2 \rangle$ optical = 0.114(9), $\delta\langle r^2 \rangle$ electronic x-ray (optically smoothed) = 0.106(7), and $\delta\langle r^2 \rangle$ muonic x-ray = 0.106 fm^2. (Optical smoothing makes no difference.)

The values of $\delta\langle r^2 \rangle$ that can be obtained from the isotope shifts given by Bonn *et al.* (1976), Kühl *et al.* (1977), and Dabkiewicz *et al.* (1979), with the aid of Eq. (9.17), are shown in Fig. 9.10 and are discussed in Section 10.2.

9.81. THALLIUM

The ground state of Tl I is $6s^2 6p\ ^2P_{1/2}$ and the excited states have a single excited outer electron outside $6s^2$. Isotope shifts in Tl I owe more to the change during a transition of the screening of the $6s^2$ electrons than to the single outer electron. The situation is very similar in In I (see Section 9.49). Not very much work has been done on the screening effects in thallium and it

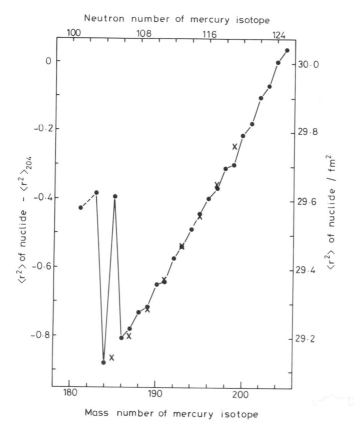

FIGURE 9.10. Mean square radii of the nuclear charge distribution of mercury isotopes. Points labeled (●) are ground state nuclides; points labeled (×) are isomers. The changes in $\langle r^2 \rangle$ from $\langle r^2 \rangle$ of ^{204}Hg are more precisely values of $\lambda/0.93$ as explained in the text.

is not possible to determine reliable values of the change in the nuclear charge radius from the measured isotope shifts, which are of course predominantly field shifts. There are two stable isotopes, ^{203}Tl and ^{205}Tl, and recent spectroscopic work has been concerned at least as much with the magnetic hyperfine structure as with the isotope shift (Flusberg *et al.*, 1976; Garton *et al.*, 1974).

Even if absolute values of $\delta\langle r^2 \rangle$ cannot be obtained, it is of interest to compare relative isotope shifts for neighboring elements. Goorvitch *et al.* (1969) have increased the amount of overlap with mercury by studying neutron-deficient isotopes produced by the alpha bombardment of ^{197}Au foil. The thallium isotopes were preferentially produced in high angular momentum isomeric states so isomer as well as isotope shifts were measured. Their work showed that from $N = 118$ to $N = 124$, the relative isotope shifts of the two neighboring elements are very similar for the same neutron numbers. For isotopes of neutron number less than 118 the similarity is less evident. The heavy nuclides are nearer the magic neutron number $N = 126$ and so the effect of the single odd nucleon is more dominant than it is in the lighter nuclides where individual nucleons and their interactions come into play.

9.82. LEAD

Lead is the heaviest element with more than one stable isotope. It owes its stability to the fact that it has a magic number of protons and that one of its isotopes, ^{208}Pb, also has a magic number of neutrons. Isotope shifts have been measured in the resonance line $6s^2 6p^2\,^3P_0 - 6s^2 6p7s\,^3P_1$ by Thompson et al. (1983). The interpretation of the shifts in terms of nuclear properties is difficult because the change in the screening of $6s^2$ during the transition is significant and difficult to estimate. Thompson et al. (1983) avoided the problem by comparing their isotope shifts with muonic x-ray energy and electron scattering studies. They believe that there is strong evidence for the specific mass shift being as large as ten times the normal mass shift. This means the k factor could be as large as 1.6 a.u. which, if true, would certainly be unique. This uncertainty in the mass shifts makes very little difference to the interpretation of the isotope shifts in terms of nuclear charge parameters. The field shifts are proportional to λ of Eq. (5.4). Using the coefficients tabulated by Seltzer (1969)

$$\lambda = \delta\langle r^2\rangle - 1.1\times 10^{-3}\delta\langle r^4\rangle + 3.0\times 10^{-6}\delta\langle r^6\rangle \qquad (9.18)$$

Thompson et al. (1983) estimated that for reasonable charge distribution in lead nuclei without deformation

$$\lambda = \delta\langle r^2\rangle(1 - 0.084 + 0.014) = 0.93\delta\langle r^2\rangle \qquad (9.19)$$

Fifteen isotopes, including many radioactive ones, were studied. In addition to significant odd–even staggering, there is a clear discontinuity in the trend of $\delta\langle r^2\rangle$ on either side of ^{208}Pb. As would be expected, the field shifts are much larger for isotopes heavier than ^{208}Pb. From ^{198}Pb to ^{208}Pb, $\delta\langle r^2\rangle$ for $\delta N = 2$ is 0.11 fm^2; from ^{208}Pb to ^{212}Pb it is 0.20 fm^2.

Muonic x-ray energies were measured by Kessler et al. (1975) and analyzed by Ford and Rinker (1973) in a model-independent way. More recently, muonic x-ray energies of isomers have been measured (see Section 7.3.2 for the references).

TABLE 9.21

Isotope Shifts and Changes in Nuclear Charge Radii for Bismuth Isotopes

Mass numbers of isotope pair	Isotope shift (MHz)[a]	$\beta\lambda(\text{fm}^2)$ [b]
209, 207	819(150)[c]	0.079(22)
209, 206	3630(390)[d]	0.350(57)
209, 205	5760(420)[e]	0.555(61)

[a] In $\lambda 306.7$ nm, $6p^3\,{}^4S_{3/2}-6p^27s\,{}^4P_{1/2}$.
[b] Using the analysis of Heilig and Steudel (1974).
[c] Chuckrow and Stroke (1971).
[d] Magnante and Stroke (1969).
[e] Mariño et al. (1975).

9.83. BISMUTH

Bismuth was one of the first elements in which muonic x-ray energies were measured. Bardin et al. (1967) studied the one stable isotope, ^{209}Bi, and found that $\langle r^2 \rangle$ is 30.40(73) fm^2. Optical isotope shifts have been measured between ^{209}Bi and ^{207}Bi, ^{206}Bi, and ^{205}Bi. The results are given in Table 9.21 where λ is approximately $0.93\ \delta\langle r^2 \rangle$ as in lead. The screening factor β, which arises mainly from the change in the screening of the $6s^2$ electrons during the transition, has not been evaluated.

9.84. POLONIUM

Charles (1966) used isotope shifts as an aid in the extension of the analysis of the spectrum.

9.85. ASTATINE

There is nothing to report.

9.86. RADON

There is nothing to report.

9.87. FRANCIUM

Francium is an alkali metal, and yet not a line of its simple spectrum was observed until 1978. The very simplicity of its spectrum made it impossible to

TABLE 9.22

Isotope Shifts in λ717.97 nm ($7s\,^2S_{1/2}-7p\,^2P_{3/2}$) of Francium[a]

Mass number	208	209	210	211	212	213
Shift (MHz)	6645(9)	4771(8)	4243(8)	2534(8)	1628(8)	0

[a] Figures from Liberman *et al.* (1980).

disentangle it from the complicated spectra of the elements with which it was always associated. Liberman *et al.* (1980) studied the spectrum of chemically and isotopically pure francium isotopes produced at the CERN ISOLDE on-line mass separator by means of the spallation of uranium by 600-MeV protons. This produced about 10^8 atoms per second; enough for the $7s\,^2S_{1/2}-7p\,^2P_{3/2}$ transition at 717.97 nm to be detected by laser atomic beam spectroscopy as used for other alkali metals and described by Huber *et al.* (1978). It was possible to measure its hyperfine structure and the isotope shifts given in Table 9.22. Work is in progress to find the other D line and to measure the nuclear moments. When this is done, it will be possible to deduce values of the change between isotopes in the charge distribution of the nucleus.

9.88. RADIUM

There is nothing to report.

9.89. THE ACTINIDES

Isotope shift measurements have contributed greatly to the analysis of the complex spectra of the actinides. The isotope shifts are predominantly field shifts and so are proportional to the change in the electron density at the nucleus. Wilson (1968) included plutonium in his evaluations of $|\psi(0)|^2$ for electron configurations in heavy atoms, but Rajnak and Fred (1977) have shown that better agreement with experimental shifts is obtained if the pseudorelativistic Hartree–Fock model of Cowan and Griffin (1976) is used (see Section 4.2.2). The agreement can be shown in diagrams such as Fig. 9.11 where measured isotope shifts between fairly pure configurations are plotted against evaluations of $4\pi|\psi(0)|^2 a_0^3$ for these configurations. The amount of data to be considered is enormous since the actinides have complex spectra and much configuration mixing. More work has probably been done on uranium than on any other actinide, so let us consider it first. The enormous amount of data can be indicated by quoting from Blaise and Radziemski

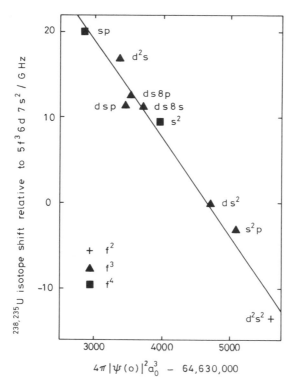

FIGURE 9.11. A comparison of isotope shifts in U I (Blaise and Radziemski, 1976) with the relativistic evaluations of $4\pi|\psi(0)|^2 a_0^3$ (Rajnak and Fred, 1977) for various configurations.

(1976), who mention: "92,000 spectral lines", "35,000 classified lines", "5,500 isotope shifts measured for ^{238,235}U", "1,600 known levels", and give their estimate of the isotope shift for the lowest levels of 12 configurations of U I. The shifts of nine of these levels are plotted against the evaluations of $4\pi|\psi(0)|^2 a_0^3$ in Fig. 9.11 to show how good the overall agreement is. For three other levels the agreement is not so good, but as Rajnak and Fred (1977) point out, the deviations are always in a direction which can be explained quite naturally by an appeal to the effects of configuration mixing. With the aid of this diagram, isotope shifts can be predicted in configurations which have not yet been identified. This will help in their identification, and so aid the further analysis of the first spectrum of uranium. Engleman and Palmer (1980) have also measured isotope shifts in thousands of lines using Fourier transform spectroscopy.

Gagné et al. (1976; 1978) have concentrated on just six spectral lines and measured the isotope shifts between more than one pair of isotopes. This allows the relative field shifts and the relationship between the mass shifts to

TABLE 9.23

Isotope Shifts in U I

Wavelength (nm)[a]	Lower level	Upper level (cm^{-1})	Relative field shift	Mass shift ^{238,236}U (GHz)[b]
578.059	$^7M_6$	23547	$+0.96(4)$	$-0.96\,m$
516.415	$^7M_7$	27478	$+0.60(1)$	$-0.60\,m$
491.035	$^7M_6$	26607	$+0.96(2)$	$-0.57-0.96\,m$
506.377	$^5L_7$	23547	$-1.00(1)$	m
502.738	$^5L_6$	19887	$-1.54(3)$	$0.51+1.54\,m$
501.142	$^5K_5$	20564	$-1.39(2)$	$0.54+1.39\,m$
Pseudotransition				
506–578	$^5L_7$	$^7M_6$	$-1.96(4)$	$1.96\,m$

[a] Isotope shifts measured by Gagné et al. (1976; 1978).

[b] The value of m cannot be found from the experimental data, but the mass shift in the pseudotransition would be expected to be small as only the f^3ds^2 and f^3d^2s configurations are involved. The normal mass shift is 0.01 GHz in the real transitions.

be found from a King plot. The more recent work on ^{233}U enables much more precise values to be obtained than those given by King (1979) and these are given in Table 9.23. The mass shifts have a large uncertainty, because of the extrapolations involved on the King plot, of about 0.15 GHz in each spectral line. The mass shifts are given in terms of the mass shift in λ506 nm which is probably small because the mass shift in the pseudotransition 506 nm–578 nm is expected to be small since there is no change in the number of 5f electrons. Bauche (1974) has made a Hartree–Fock evaluation of the specific mass shift of a 5f electron and found it to be 0.3 GHz. The negative mass shift in λ491 nm suggests that there is an admixture of a $4f^2$ configuration in its upper level. The positive mass shifts in λ503 nm and λ501 nm suggest an admixture of a $4f^4$ configuration in their upper levels.

Heilig and Steudel (1974) determined the changes in the mean square radius of the nuclear charge distribution from the isotope shift constant values of Blaise and Steudel (1968). For ^{238,235}U they found $\beta\delta\langle r^2\rangle = 0.364$ fm^2 based on shifts measured in U I and U II. From the evaluations of Rajnak and Fred (1977), the screening factors β in the transitions can be determined and give $\delta\langle r^2\rangle = 0.340$ fm^2, but with big differences between the results from different transitions.

Close et al. (1978) have deduced nuclear charge distributions from their measurements of muonic x-ray energies. They assumed the nuclear charge was a deformed Fermi distribution and obtained the values given in Table 9.24. Due to changes in size and shape, $\delta\langle r^2\rangle$ is estimated from these to be about 0.29 fm^2 for ^{238,235}U, a result which is somewhat lower than the value deduced from optical isotope shift work. The relative field shifts (and hence

TABLE 9.24

Nuclear Parameters of Uranium

	a(fm)[a]	c(fm)[a]	$\beta_2{}^a$	$\langle r^2 \rangle_{\text{sph}}$(fm)[2 b]	$\langle r^2 \rangle$(fm)[2 c]
^{235}U	0.454(6)	7.043(8)	0.272(2)	32.61	33.95
^{238}U	0.448(4)	7.076(6)	0.279(2)	32.82	34.24

[a] Values of a, c and β_2 deduced by Close *et al.* (1978) from their muonic x-ray energy and hyperfine structure measurements.
[b] $\langle r^2 \rangle_{\text{sph}} = \frac{3}{5}c^2 + \frac{7}{5}\pi^2 a^2$ as in Eq. (4.4).
[c] Assuming that $\langle r^2 \rangle = [1 + (7/4\pi)\beta_2^2]\langle r^2 \rangle_{\text{sph}}$ by analogy with Eq. (4.40).

relative values of $\delta\langle r^2 \rangle$), based on measurements made by Gagné *et al.* (1976; 1977; 1978) are as follows: 238,236 1.00; 238,235 1.67; 238,234 1.99; 238,233 2.59.

The other actinide elements can be dealt with more briefly. Heilig and Steudel (1974) deduced that $\beta\delta\langle r^2 \rangle = 0.205(33)$ fm^2 for ^{232,230}Th from the isotope shift constant determined by Blaise and Steudel (1968). The relative field shifts are: 232,230 1.00; 230,229 0.74. Close *et al.* (1978) deduced from their muonic x-ray measurements that, for ^{232}Th, $a = 0.449(4)$ fm and $c = 7.024(6)$ fm in a deformed Fermi distribution.

Tomkins and Gerstenkorn (1967) measured isotope shifts in several lines of the plutonium spectrum for five isotopes. Their measurements were sufficiently precise to show evidence of mass shifts of about 0.3 GHz, presumably specific mass shifts in transitions where the number of 5f electrons changed. The values of $\delta\langle r^2 \rangle/$fm^2 for six isotopes of plutonium were given by Heilig and Steudel (1974): ^{244,242}Pu, 0.139(21); ^{242,240}Pu, 0.138(21); ^{240,238}Pu, 0.185(28); ^{241,240}Pu, 0.052(10); ^{241,239}Pu, 0.161(27); and ^{239,238}Pu, 0.074(11). Recent work on isotope shifts in plutonium has yet to be published. Close *et al.* (1978) deduced from their muonic x-ray measurements that, for ^{230}Pu, $a = 0.447(14)$ fm and $c = 7.091(16)$ fm in a deformed Fermi distribution.

In the heavier actinides there is expected to be less configuration mixing, and the isotope shifts measured in americium by Fred and Tomkins (1957) agree well with the evaluations of $|\psi(0)|^2$ by Rajnak and Fred (1977). Blaise and Steudel (1968) used three methods to find the isotope shift constant and they gave consistent results. Heilig and Steudel (1974) deduced from this work that $\beta\delta\langle r^2 \rangle = 0.147(25)$ fm^2 for ^{243,241}Am. The evaluations of Rajnak and Fred show that $\beta = 1.03(2)$ for the transitions involved, and so $\delta\langle r^2 \rangle = 0.143(25)$ fm^2. Bemis *et al.* (1979) measured the isomer shift for ^{240}Am in an optical transition by the use of a laser-excited optical pumping technique. The shift is huge, implying that $\delta\langle r^2 \rangle = 5.1(2)$ fm^2 between the spontaneously fissioning isomer and the ground state. Whatever might be argued about possible odd–even staggering between the americium isotopes, this isomeric $\delta\langle r^2 \rangle$ shows that the isomer must have a huge deformation. Bemis *et al.*

(1979) deduced that $\beta_2 = 0.66(4)$ for the spontaneously fissioning isomer as compared with 0.24 for americium nuclei in their ground state. The largest deformation in the rare earth region is about 0.3, so 0.66 is quite uniquely large.

Worden and Conway (1976) measured isotope shifts in many lines of the first spectrum of curium. They are in reasonable agreement with the evaluations of $|\psi(0)|^2$ by Rajnak and Fred (1977). More recently, the relative field shifts have been measured: 246,244 1.00; 245,244 0.35(2); 248,244 1.95(1). Some recent work on isotope shifts in curium, carried out primarily as an aid in spectral analyses, has yet to be published.

APPLICATIONS AND CONCLUSIONS

10.1. ATOMIC ELECTRON STRUCTURE FROM ISOTOPE SHIFTS

The information which can be obtained from isotope shifts is very different in diffcrent elements. In the simplest element, hydrogen, the interest is more in the shift from the hypothetical infinitely heavy point nucleus than in the shift between actual isotopes. This is because there is no doubt about the structure of a single-electron atom but a comparison of experiment with theory enables the theory involved, relativistic quantum electrodynamics, to be tested. One quantity which at present has to be inserted as an experimental fact of life, without any generally accepted evaluation of its value from theory, is the electron-to-proton mass ratio. As mentioned in Section 3.1, the best value for this ratio may come from isotope shift measurements in the spectrum of hydrogen, but an alternative method (Gräff et al., 1980), which may give even higher precision, is to compare the cyclotron frequencies of protons and electrons in the same magnetic field. Another quantity which has to be measured is the size of the nuclear charge distribution, but in the case of hydrogen the field shift is barely significant, even in the most precise work, and so only an approximate value is required.

By contrast, for the heaviest elements the field shift is the dominant part of the isotope shift. Since the size of the field shift for a particular pair of isotopes is proportional to the change in electron density at the nucleus during the transition, $\Delta|\psi(0)|^2$, comparison of isotope shifts in different transitions gives a measure of how $\Delta|\psi(0)|^2$ varies between transitions. Transitions in which the number of s electrons change give large values of $\Delta|\psi(0)|^2$, and so the comparison of isotope shifts in different lines of a spectrum can be a great help in the interpretation of a spectrum in which the electron configurations are unknown. This technique has proved particularly useful in the very complicated spectra of the rare earths and of the actinides. With pure configurations the change in the number of s electrons must be integral, but when configuration mixing is present then intermediate values are also possible. The screening effects of other electrons can also help to smear out the field shifts into a continuum of values. A good example of the use of field shifts to determine the amount of configuration mixing that is present is the work of Champeau (1972) on Ce I and Ce II. Isotope shifts played an important role in the interpretation of Nd I; Rao and Gluck (1964), for instance, were able to classify quite a bit of the spectrum with the aid of the isotope shifts they had measured in about 50 spectral lines. Recent examples of where the measurement of isotope shifts has helped in the analysis of spectra are the work of Ahmad and Saksena (1981) on neodymium, of Aufmuth (1978) on dysprosium, of Miller and Ross (1976) on erbium, of Buchholz *et al.* (1979) on rhenium, and of Grethen *et al.* (1980) on platinum.

In the case of the actinides, the main reason for measuring isotope shifts has been to help in the analysis of the spectra. This topic is mentioned in Section 9.89, and the most useful general reference that can be mentioned here is the work of Rajnak and Fred (1977) who have shown the need to use relativistic evaluations for pure configurations when deducing the amount of configuration mixing from measured isotope shifts. In the spectra of these very heavy elements the isotope shifts are predominantly field shifts, but for lighter elements it is possible, in principle, to get more information by determining the mass shifts as well as the field shifts. This could be an additional aid in determining electron configurations since large mass shifts are not caused by the same electron configurations as are large field shifts. In practice little use has been made of this technique because shifts have to be measured between more than one pair of isotopes before the mass shifts can be separated from the field shifts. Perhaps the most thorough example of such an analysis is that on Xe I made by Jackson *et al.* (1975). The mass and field shifts of 30 levels were described by six parameters, three of which were involved in first-order field shifts, five in second-order field shifts, and all six in first-order mass shifts.

It is surprising that so much about atomic electrons can be deduced from their interaction with the tiny nucleus at the center of the atom. It is particularly surprising that this interaction is detectable for electrons in highly excited Rydberg levels. Niemax and Pendrill (1980) measured the isotope shifts for ^{41,39}K in 4^2S$-n^2$S and 4^2S$-n^2$D transitions, and were able to deduce the shifts of n^2S ($7 \leq n \leq 18$) and n^2D ($5 \leq n \leq 17$) levels relative to the ionization limit $3p^6\,^1S_0$. Their results are given in Table 10.1 and interpreted in the light of the shifts measured in 4^2S-4^2P by Touchard *et al.* (1982a). It can be seen that there are significant variations of the specific mass shift with the principal quantum number, even up to $n = 8$.

Isotope shifts have also revealed a lot about the screening effects of transition electrons on other electrons in the atom. Screening ratios can be measured experimentally and compared with evaluations from theory [see Eq. (4.34) of Section 4.2.2]. The agreement between experiment and theory is good enough to encourage the evaluation from theory of screening factors [see Eq. (4.35)], which cannot be measured experimentally. These are required in order to determine the change between isotopes in the mean square radius of the nuclear charge distribution, $\delta \langle r^2 \rangle$, from the field shift. Values of $\delta \langle r^2 \rangle$ determined in this way are not always in good agreement with values

TABLE 10.1
Isotope Shifts for ^{41,39}K/MHz between Rydberg Levels[a]

Lower level[b]	SMS + FS	Probable SMS[c]
4^2S	−61(2)	−46
7^2S	−32(8)	−32
8^2S	−15(8)	−15
9^2S	−4(8)	−4
≥ 10^2S	0(9)	0
5^2D	−53(8)	−53
6^2D	−36(10)	−36
7^2D	−11(9)	−11
8^2D	−6(9)	−6
9^2D	−11(10)	−11
≥ 10^2D	0(9)	0

[a] Figures from Niemax and Pendrill (1980).
[b] In all cases the upper level of the transition is the ionization limit $3p^6\,^1S_0$.
[c] Assuming the field shift (*FS*) in 4S–∞ is the same as in 4S–4P as determined by Touchard *et al.* (1982a) and that all other field shifts are negligible.

determined from the measurement of muonic x-ray energies. Cases where the discrepancies are worryingly large, being on the order of 20%, are in the spectra of nickel, molybdenum, ytterbium, osmium, and uranium. The problem may lie in our lack of understanding of mutual electron screening, but wheresoever it does lie, it is a problem that must be resolved if the subject is to proceed satisfactorily in the future.

10.2. NUCLEAR STRUCTURE FROM ISOTOPE SHIFTS

The overall picture is very conveniently displayed in a Brix and Kopfermann diagram in which something related to the field shift is plotted against the neutron number of the heavier isotope of the pair. The neutron number is used because field shifts are much more dependent on the neutron number than on the proton number, as can be seen from the scatter of the points in Fig. 10.1. Except for $N = 22$ and $N = 90$, there is not much scatter between the points for different elements at the same value of the neutron number N. It was traditional to plot the observed isotope shift constant divided by that for the simplest possible model of the nucleus based on Eq. (4.1) [see, for example, Brix and Kopfermann (1958)], but nowadays, as has been pointed out by Gerstenkorn (1973), it is more natural to plot the change between isotopes in the mean square radius of the nuclear charge distribution as the ordinate. This has been done in Fig. 10.1. On the basis of Eqs. (4.1) and (4.3)

$$\delta\langle r^2\rangle = 0.58A^{-1/3}\delta A \tag{10.1}$$

and the curve corresponding to this simplest of nuclear models is drawn in Fig. 10.1. Each point comes from an isotope shift measurement, usually an optical one but in some cases a muonic one, involving only even neutron numbers so as to avoid odd–even staggering effects.

It is quite clear that, although Eq. (4.1) describes approximately correctly the sizes of the stable nuclides distributed across the Periodic Table, its derivative form of Eq. (10.1) overestimates the increase in nuclear size between isotopes. Only 20 of the 126 points lie above the curve of Eq. (10.1). This is only to be expected as the former is concerned with the addition of

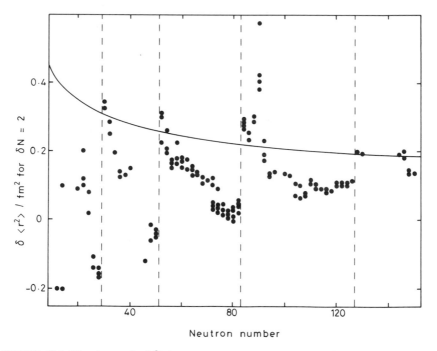

FIGURE 10.1. The changes in $\langle r^2\rangle$ between isotopes plotted against the neutron number of the heavier isotope. All the results are for even neutron numbers and involve a change of two in the neutron number. The plotted points are for stable isotopes or those that lie close to the region of stability on the Segrè chart. Different elements are not distinguished but all the individual results are to be found in the relevant sections of Chapter 9. The curve is obtained from Eq. (10.1) which is based on the simplest possible nuclear model of Eq. (4.1). The discontinuities shown by vertical dashed lines occur at magic numbers.

neutrons and protons in almost equal numbers, whereas the latter is concerned with the addition of neutrons alone. Nevertheless, there is, in the great majority of cases, an increase in $\langle r^2 \rangle$ between isotopes; in general, the extra neutrons of the heavier isotope cause it to have a larger nuclear charge distribution. This shows that the neutrons and protons in nuclei interact as is to be expected on the liquid drop model. It is also noticeable that the experimental values of $\delta\langle r^2 \rangle$ tend to the liquid drop model value of Eq. (10.1) as the number of nucleons increases.

But perhaps the most obvious features of Fig. 10.1 are the large discontinuities that occur at the four neutron numbers indicated by vertical dashed lines: 29, 51, 83, and 127. Since the neutron number that is plotted is that of the heavier isotope of the pair, it follows that nuclides with $N = 28$, 50, 82, and 126 must have small values of $\langle r^2 \rangle$ when compared with neighboring isotopes. These numbers are, of course, some of the so-called "magic numbers," and the shell model explains why these nuclei are particularly compact: it is because they consist entirely of closed shells of neutrons. Extra neutrons added to closed shells are relatively loosely bound and so $\delta\langle r^2 \rangle$ is large. As more neutrons are added, they are more and more tightly bound as the next magic number is approached. This can obviously be seen to be happening from $N = 28$ to $N = 50$ and from $N = 50$ to $N = 82$, but beyond $N = 82$ there is an additional phenomenon. From $N = 84$ onward $\delta\langle r^2 \rangle$ does not decrease but rises to a maximum value at $N = 90$. The uniquely large value of $\delta\langle r^2 \rangle$ is for [153,151]Eu. These large shifts are due to nuclear deformation as was pointed out by Brix and Kopfermann (1949) and as was explained in Section 4.2.3. At $N = 90$, not only is the quadrupole deformation parameter β_2 fairly large, but $\delta\beta_2/\delta N$ is also large and positive, giving a large shape component to $\delta\langle r^2 \rangle$.

The size of nuclei is related to their nuclear binding energies. To take an extreme example, a nuclide with a magic number of neutrons has a large binding energy and is small compared with neighboring nuclides. The correlation between isotope shifts and changes between isotopes in the nuclear binding energy per nucleon has been elaborated by Gerstenkorn (1971; 1973) and by others (Wenz et al., 1980).

Let us now leave the overall picture and look at some of the details of nuclear structure that are revealed by the measurement of isotope shifts. Perhaps the lightest element that is worthy of mention is carbon. Both electron scattering experiments and muonic x-ray isotope shifts (Schaller et al., 1982) suggest that $\langle r^2 \rangle$ increases much more between ^{13}C and ^{14}C than it does between ^{12}C and [13]C. The reason for this is not obvious, since ^{12}C has a closed $1p_{3/2}$ neutron subshell and the additional neutrons in ^{13}C and ^{14}C both go to fill up the $1p_{1/2}$ subshell. Schaller et al. (1982) point out that the explanation does not lie in a larger deformation in ^{14}C than ^{12}C either, and conclude that "there are still open questions regarding the structure of the carbon isotopes."

Sodium is the lightest element in which isotope shifts have been measured for a long series of radioactive isotopes (Touchard et al., 1982b). The obvious feature of their results is the small changes in $\langle r^2 \rangle$ up to ^{25}Na, and the fairly large increases in $\langle r^2 \rangle$ between ^{25}Na and the heavier isotopes. This is obviously a shell effect. ^{25}Na has a closed $1d_{5/2}$ neutron subshell and the additional neutrons in ^{26}Na to ^{31}Na go to fill up the $1d_{3/2}$ and $2s_{1/2}$ subshells. The absolute values of $\delta\langle r^2 \rangle$ can only be estimated because it is not known exactly how the isotope shift is divided between the mass shift and field shift. Hartree–Fock calculations of the distribution of the nuclear charge suggest that the change in deformation between ^{25}Na and ^{27}Na is very small and so Touchard et al. (1982b) assumed that the field shift ^{27,25}Na was one-half of that given by Eq. (10.1). The factor one-half, sometimes called the isotope shift discrepancy, is found from the average value of the points in Fig. 10.1 divided by the value of Eq. (10.1) for the relevant A value. So with $\delta\langle r^2 \rangle = 0.20$ fm^2 for ^{27,25}Na and the use of Eq. (4.24) and Eq. (4.32) it follows that the field shift is -6 MHz and the mass shift 1140 MHz. Having estimated the mass shift, the field shifts and values of $\delta\langle r^2 \rangle$ between other pairs of isotopes can be calculated. Touchard et al. (1982b) argued that the variation in $\delta\langle r^2 \rangle$ between different isotope pairs could not be explained as a deformation effect but that the actual increase in volume on adding a $1d_{5/2}$ neutron is less than the increase on adding a $1d_{3/2}$ neutron or a $2s_{1/2}$ neutron.

Isotope shifts have been measured for a long series of isotopes of potassium (Touchard et al., 1982a) where, as with sodium, the mass shifts are much larger than the field shifts. It is, however, possible to be more quantitative in the case of potassium, since $\delta\langle r^2 \rangle$ for ^{41,39}K is known from measurements of the energies of muonic x-rays. $\langle r^2 \rangle$ increases at about 0.1 fm^2 for $\delta A = 2$ up to ^{41}K; it then increases at about 0.03 fm^2 for $\delta A = 2$ up to ^{45}K; and, finally, it decreases at about the same rate as far as ^{47}K, which is the heaviest isotope which was studied. Superimposed on these trends there is a small odd–even staggering effect for every odd-N isotope. There are no filled subshells at ^{41}K and ^{45}K, but a general decrease in $\delta\langle r^2 \rangle$ between ^{38}K and ^{47}K is to be expected since through this range of isotopes the additional neutrons (20 to 28) are in the $1f_{7/2}$ subshell. In fact, as the last two neutrons are added to fill the subshell, the radius of the nuclear charge distribution is reduced. The same thing happens in calcium, but here the reduction in size is more marked (Andl et al., 1982). The reduction in $\langle r^2 \rangle$ from ^{44}Ca to ^{48}Ca is in fact so large that, as already mentioned in Section 9.20, $\delta\langle r^2 \rangle$ for ^{40,48}Ca is approximately zero. The comparison of potassium with calcium shows another way in which calcium is more extreme than potassium, and that is in its much larger odd–even staggering which is shown in Fig. 9.2. Calcium has a magic number of protons, and the absence of one $2s_{1/2}$ proton in potassium smooths the effect which the $1f_{7/2}$ neutrons have on the nuclear charge distribution.

The quadrupole deformation parameters of the calcium isotopes have been found by Coulomb excitation experiments and are listed by Träger (1981). Using Eq. (4.42), the shape component of $\delta\langle r^2\rangle$ for ^{44,40}Ca can be found to be 0.20 fm^2 since $\delta\langle\beta_2^2\rangle = 0.04$. It can be seen from Fig. 9.2 that this is a large fraction of the total $\delta\langle r^2\rangle$ for ^{44,40}Ca. In fact, as Träger (1981) points out, all the calcium field shifts are predominantly shape effects and so all the calcium isotopes have about the same volume of nuclear charge. For recent theoretical evaluations of the charge distribution in calcium nuclei, see Brown *et al.* (1979) and Hodgson (1981), the latter being a general review of the subject of nuclear charge and matter distributions.

In the region from $N = 28$ to $N = 40$ the isotope shifts which have been measured—and they are mainly muonic x-ray shifts—show that $\delta\langle r^2\rangle$ decreases rapidly as N increases but is virtually independent of Z. This can be seen in Fig. 10.1 since the points in this region are for several different elements. This trend was noted by Shera *et al.* (1976) who also pointed out that the independence of Z is surprising since one of the elements concerned is nickel, which has a magic number of protons. They suggested that this independence of the proton configuration arises because the added neutrons interact with the entire proton core rather than with just the valence protons. Wohlfahrt *et al.* (1980) showed that theoretical evaluations are in satisfactory agreement with the experimental results so long as static deformations and also zero point quadrupole oscillations of the nuclear surface are taken into account.

As mentioned in Section 7.4.1, there are three regions on the Segrè chart near the valley of stability where nuclei have permanent deformations rather than the vibrational deformations mentioned above in the case of calcium. The lightest region of the three occurs where both proton and neutron numbers lie well away from, and in between, the magic numbers 28 and 50. The isotope shifts measured by Thibault *et al.* (1981a) in the D$_2$ line of rubidium give clear evidence of this deformation region. The field shifts between ^{76}Rb and ^{87}Rb, the region from $N = 39$ to $N = 50$, show that $\langle r^2\rangle$ is decreasing in general throughout this long region. The obvious explanation, that $\sim^{77}$Rb nuclides have a large permanent deformation, is confirmed by the tendency to a rotational structure in their excited nuclear levels. Thibault *et al.* (1981a) measured hyperfine structure as well as isotope shifts, and hence obtained values of the spectroscopic quadrupole moments, Q_I. Applying their values to the projection formula of Eq. (7.8) they found Q and hence, with the aid of Eq. (4.43), β_2. The results were very large, $\beta_2 = 0.45$ and 0.5 for ^{77}Rb and ^{76}Rb, respectively, suggesting that it was indeed valid to use Eq. (7.8) and that these nuclides are very deformed rotational nuclei.

The heaviest isotope that was studied was ^{98}Rb and the field shift for ^{97,96}Rb was found to be particularly large. To put it into the context of Fig. 10.1, $\delta\langle r^2\rangle$ for ^{97,95}Rb ($N = 60$ in Fig. 10.1) is 0.66. The 1g$_{7/2}$ subshell is closed at $N = 58$, but the shift from $N = 58$ to $N = 60$ is so large that the explanation cannot be a simple shell effect; there must be considerable permanent rotational nuclear deformation. This is not unexpected because both proton and neutron numbers are well away from magic numbers. In this case N is well above 50, but nowhere near approaching 82. These points are not plotted on Fig. 10.1 because the isotopes concerned lie well away from the bottom of the valley of stability. Self-consistent mean-field calculations have been made (Campi and Epherre, 1980) of the structure of the nuclear ground states of the rubidium isotopes and many of the features revealed by the isotope shift work were obtained.

An obvious closed-shell effect was also apparent in the field shifts, which were much smaller for $N < 50$ than they were for $N > 50$. A similar discontinuity is apparent in the case of strontium, as can be seen in Fig. 9.3. Many isotope shifts have been measured in the region from $N = 50$ to $N = 74$ and from $Z = 40$ to $Z = 50$. The results are collected in Table 9.6 and are an interesting contrast to the results in the $N = 28$ to $N = 40$ region. In both cases the values of $\delta\langle r^2 \rangle$ depend more on N than on Z, but in this region the variation with N is not monotonic and some variation with Z is apparent. At one time it was thought that the variation in $\delta\langle r^2 \rangle$ between isotope pairs was entirely due to the changes in deformation between isotopes, but King *et al.* (1966) showed that, even if the shape component of $\delta\langle r^2 \rangle$ is removed from results such as those in Table 9.6, the remaining volume components do not vary monotonically with N within a range of isotopes. A more obvious pattern in the variation of $\delta\langle r^2 \rangle$ with Z and N is revealed if $\delta\langle r^2 \rangle$ between even–even isotopes is plotted against $N - Z$ (Kuhn *et al.*, 1975). The physical significance of this is that $N - Z$ is the number of neutrons which cannot be in α-particle structures.

The isotopes of tin are an example of where optical and muonic x-ray isotope shifts can be compared to obtain more detailed information about nuclear charge distributions. The former measures $\delta\langle r^2 \rangle$ which is smaller for ^{124,122}Sn than it is for ^{122,120}Sn by about 10%, whereas the latter, in the case of 1s–2p transitions, measures $\delta\langle r^{2.26}\exp(-0.12r)\rangle$, which is the same for ^{124,122}Sn and ^{122,120}Sn. Silver and Stacey (1973b) showed that, on a Fermi distribution model, this evidence suggests that the skin is significantly thinner in ^{124}Sn than it is in the other even isotopes of tin. Some recent electron scattering experiments have been done with sufficient precision to show the difference in the nuclear charge distributions of ^{116}Sn and ^{124}Sn (Cavedon *et al.*, 1982), but these do not indicate a particularly thin skin in ^{124}Sn.

Table 9.6 is continued for heavier elements by Table 10.2 which covers the region from $N = 70$ to $N = 92$ and from $Z = 51$ to $Z = 64$. Again the values of $\delta\langle r^2 \rangle$ depend more on N than on Z. The shell closure at $N = 82$ is, of course, obvious, as it is in Fig. 10.1, but note also the minimum value of $\delta\langle r^2 \rangle$ for $N = 78, 76$. This cannot be explained on the shell model since no subshell is closed at $N = 78$, but Coulomb excitation experiments on xenon and barium show that the deformation parameter $\langle \beta_2^2 \rangle$ is about 0.02 in this region, which is large enough to cause considerable modification of the nucleon orbits from their behavior in a spherical potential. For xenon and barium the values of $\delta \langle r^2 \rangle$ are consistent with a shape component calculated from Eq. (4.42) plus a volume component of about the usual one-half of Eq. (10.1). Assuming this is also true for caesium, then the nuclear deformations of the caesium isotopes are very similar to those, for the same N, in xenon and barium (Thibault *et al.*, 1981b). Some very neutron deficient isotopes of caesium were studied for which the results are shown in Fig. 9.6 but not in Table 10.2. A very striking feature in this region is that the low-spin states $^{119m}\mathrm{Cs}$ ($I = \frac{3}{2}$), $^{121}\mathrm{Cs}$ ($I = \frac{3}{2}$) and $^{122}\mathrm{Cs}$ ($I = 1$) follow the general trend of Fig. 9.6, whereas the high-spin states $^{119}\mathrm{Cs}$ ($I = \frac{9}{2}$), $^{120}\mathrm{Cs}$ ($I = 2$), $^{121m}\mathrm{Cs}$ ($I = \frac{9}{2}$), and $^{122m}\mathrm{Cs}$ ($I = 8$) all lie higher. Attributing these isomer shifts to changes in deformation gave results that were consistent with the quadrupole moments determined from the hyperfine structure by Thibault *et al.* (1981b).

TABLE 10.2

Changes in the Mean Square Radius of the Nuclear Charge Distribution[a]

Isotope pair by neutron number	$\delta \langle r^2 \rangle$ (fm^2)								
	$_{51}\mathrm{Sb}$	$_{54}\mathrm{Xe}$	$_{55}\mathrm{Cs}$	$_{56}\mathrm{Ba}$	$_{58}\mathrm{Ce}$	$_{60}\mathrm{Nd}$	$_{62}\mathrm{Sm}$	$_{63}\mathrm{Eu}$	$_{64}\mathrm{Gd}$
Even, even									
72, 70	0.12	0.041	0.053	0.028					
74, 72		0.036	0.042	0.021					
76, 74		0.031	0.042	0.014					
78, 76		0.022	0.014	0.010					
80, 78		0.030	0.025	0.011	-0.006				
82, 80		0.047	0.057	0.040	0.020				
84, 82			0.278	0.269	0.270	0.277			
86, 84			0.256		0.233	0.257			
88, 86			0.264			0.286	0.303		
90, 88			0.216			0.381	0.423	0.577	0.407
92, 90							0.230		0.174
Odd, even									
73, 72			-0.015	-0.011					
75, 74		-0.005	0.010	-0.010					
77, 76		-0.018	-0.023	-0.022					
79, 78		-0.016	-0.010	-0.032					
81, 80			-0.008	-0.023					
83, 82			0.120						
85, 84			0.126						
87, 86			0.130				0.092		

[a] For the sources of the data and other details see the relevant sections of Chapter 9.

From $N = 82$ the deformations increase rapidly with N and there are extremely large shape contributions to $\delta\langle r^2\rangle$. This effect is at its largest as N changes from 88 to 90, as can be seen in Table 10.2. Each element except caesium has its largest $\delta\langle r^2\rangle$ for this pair of neutron numbers. Caesium differs from the other elements because the latter have permanently deformed rotational nuclei, whereas Thibault *et al.* (1981b) have shown (from a comparison of isotope shifts with quadrupole moments) that for $N \geq 88$ the caesium nuclei are probably vibrational. The values of $\delta\langle r^2\rangle$ depend more on changes in shape than on changes in volume for neutron numbers around 90, as can be seen in Table 4.6 for the case of samarium and Table 9.14 for the case of neodymium. Such a sudden change in shape is consistent with the spectra of nuclear excited states. Nuclei with $N = 88$ have a low-lying spectrum typical of spherical nuclei, whereas nuclei with $N = 90$ have a spectrum typical of deformed nuclei. ^{152}Sm was the first even–even nuclide which was recognized to have a large deformation. Having no nuclear moment, the spectroscopic quadrupole moment is zero, but Brix and Kopfermann (1949) deduced from the large isotope shifts between ^{152}Sm and lighter isotopes that ^{152}Sm has a large intrinsic quadrupole moment because of its large deformation. ^{152}Sm in the ground state has a pure rotational character (Yamazaki *et al.*, 1978), but the isomer shifts measured in the muonic x-ray work suggest that this is not so for the isomeric 2^+ state in ^{152}Sm. By far the biggest change in $\langle r^2\rangle$ between $N = 88$ and $N = 90$—indeed it is the largest $\delta\langle r^2\rangle$ in the whole of Fig. 10.1—is in europium. This is because europium is the only element in this region of the Segrè chart with an odd number of protons for which isotope shifts have been measured. The odd proton cannot pair off with another proton and so europium is less stable and tightly packed than its even–even neighbors. This pairing effect is so important that it forms the basis of the interacting boson model of the nucleus which has recently become popular. The large change in $\langle r^2\rangle$ between $N = 88$ and $N = 90$ arises because the deformation is much larger for $N = 90$. The deformation continues to increase with N as the magic number of 82 neutrons is left behind and reaches a maximum at about $N = 100$. However, the rate of increase falls off, and so $\delta\langle r^2\rangle$ is much smaller, as can be seen in Fig. 10.1. In the region around $N = 100$, where the shape component of $\delta\langle r^2\rangle$ is negligible, the measured values of $\delta\langle r^2\rangle$ are about one-half of the value of Eq. (10.1), just as they are on average for nuclei with little or no deformation. As the next magic number of neutrons (126) is approached, the deformation is decreasing and the shape component of $\delta\langle r^2\rangle$ is negative. The measured values of $\delta\langle r^2\rangle$ do not reach such low values as when the magic number 82 is approached. They do not reach their lowest value so near to the magic number either. Approaching $N = 82$ the lowest value is at about 78, 80, whereas approaching $N = 126$ it is at about 114, 116. The change from deformed to spherical nuclei approaching $N = 126$ is clearly much more gradual than the change from

TABLE 10.3

Nuclear Spin Quantum Numbers, I, and Odd–Even Staggering
Parameters, γ, for Mercury

Mass	Ground state		Isomeric state	
number	I	γ^a	I	γ^a
185	$\frac{1}{2}$	13	$\frac{13}{2}$	0.4^b
187	$\frac{3}{2}$	0.8	$\frac{13}{2}$	0.1
189	$\frac{3}{2}$	0.4	$\frac{13}{2}$	0.2
191	$\frac{3}{2}$	0.4	$\frac{13}{2}$	0.5
193	$\frac{3}{2}$	0.6	$\frac{13}{2}$	0.8
195	$\frac{1}{2}$	1.0	$\frac{13}{2}$	0.8
197	$\frac{1}{2}$	0.7	$\frac{13}{2}$	0.9
199	$\frac{1}{2}$	0.3	$\frac{13}{2}$	1.3
201	$\frac{3}{2}$	0.6		
203	$\frac{5}{2}$	0.6		

[a] The uncertainty in γ is about 1 in the last figure except for the case of ^{185m}Hg.

[b] The uncertainty in this case is about 0.4.

spherical to deformed as $N = 82$ is left behind. There is no neutron number in Fig. 10.1 with uniquely small values of $\delta\langle r^2\rangle$ corresponding to $N = 90$ with its uniquely large values of $\delta\langle r^2\rangle$.

Another regular trend has been pointed out by Stroke *et al.* (1979) for mercury isomers with neutron numbers from 105 to 119. In this range, all the odd neutrons have $I = \frac{13}{2}$ and the odd–even staggering parameter goes from about zero to about unity. The relevant figures are collected in Table 10.3. Since the paired neutrons also go into $I = \frac{13}{2}$ orbits, one would expect the effect of one neutron to be about half that of two paired neutrons leading to a value for the odd–even staggering parameter of about unity. Stroke *et al.* (1979) argued that this is not true for the lighter isomers because the Coriolis force, which is large on a neutron with an angular momentum as large as $\frac{13}{2}$, can decouple the neutron from the core so that the neutron's angular momentum points along the axis of rotation of the deformed nucleus rather than along the axis of symmetry of the nucleus.

For neutron numbers of 105 or less there is a huge new effect, as can be seen in Fig. 9.10. In this region there are clearly two competing sets of values of $\langle r^2\rangle$. The larger one, which must arise from a larger deformation, is adopted by ^{181}Hg, ^{183}Hg, and ^{185}Hg, whereas the smaller one is adopted by ^{184}Hg, ^{185m}Hg, and ^{186}Hg. The larger deformation is in fact prolate with β_2 about $+0.3$ and the smaller deformation is oblate with β_2 about -0.15 (Dabkiewicz *et al.*, 1979). Such a change in deformation is large enough for the isomer shift of ^{185}Hg to be entirely a shape effect, for if $\langle r^2\rangle = 29$ fm^2 and $\delta(\beta_2^2) = 0.06$, then, according to Eq. (4.42), $\delta\langle r^2\rangle_{\text{shape}} = 0.7$ fm^2. The measured value of the isomer shift expressed as a change in $\langle r^2 \rangle$ is in fact 0.52 fm^2.

Theoretical evaluations of the change between isotopes in the mean square radius of the nuclear charge distribution have so far failed to give values in good agreement with the experimental values. This is not surprising when it is considered that isotope shifts give values for a rather subtle nuclear quantity, namely, the change in the change in the size of the nuclear charge distribution between different pairs of isotopes. Microscopic theories of the nucleus, particularly those that consider the nucleons to be strongly paired together and behaving as bosons (in this context IBM stands for interacting boson model) have been more successful in the prediction of isomer shifts than isotope shifts. See, for example, the paper of Yamazaki *et al.* (1978) on isotope and isomer shifts in samarium.

10.3. SEPARATION OF ISOTOPES

It was first proposed by Hartley *et al.* (1922) that the isotope shift could be used to differentiate between different isotopes and hence lead to their separation. The advent of powerful tunable lasers made this proposal a practical possibility in the 1970s. A sufficiently monochromatic laser field can excite one isotope selectively from a mixture of isotopes, the other isotopes being out of resonance and so not excited. The excited isotope can then be made to partake preferentially in some chemical process (as in the development of a photographic emulsion) or it can be preferentially ionized or dissociated by another laser field. In practice, it is the isotope shifts in molecular spectra that are used rather than those in atomic spectra.

The separation process that has enormous economic possibilities is the enrichment of uranium. A more efficient way of enriching uranium than the gaseous diffusion or centrifuge method would save huge sums of money. Much work is being done on the ground state vibrational transition of UF_6 which requires a laser beam with a wavelength of 16 μm that can select from a shift between isotopes of about 19.5 GHz. Such a laser beam has been produced by a CO_2-pumped CF_4 laser, and uranium enrichment has been demonstrated. A more economical light source is what is now needed, and a great deal of work is going on to try to find it; one possibility is the free electron laser.

Isotopic separation using the isotope shift in the atomic rather than the molecular spectrum has been carried out in a few cases. Two examples, europium and dysprosium, are mentioned in Sections 9.63 and 9.66, respectively. Both of these used charge transfer ionization to ionize the excited isotope preferentially and so deflect it from the atomic beam. A very elegant, but unfortunately inefficient, way of deflecting one isotope from an atomic beam is to cross a laser beam perpendicularly with an atomic beam. The isotope that is selectively excited by the laser beam receives momentum from the laser beam but spontaneously reemits isotropically. The net force is in the direction of the laser beam, and so the isotope is deflected out of the atomic beam as a result of the multiple absorption and emission processes that occur as the isotope crosses the laser beam.

10.4. CONCLUDING REMARKS

The present healthy state of interest in the subject of isotope shifts can be judged from Fig. 10.2, which shows that the number of references cited in this book increases as the present date is approached. The exponential increase with time is not, of course, the increase in the total number of papers on isotope shifts and related topics; many such papers which have been superseded by more modern work are not cited. The exponentially increasing number of papers that are currently significant does, however, indicate that the subject is in a healthy state, and is showing no significant sign of decline. The two emission features and the one absorption feature of the spectrum of Fig. 10.2 are perhaps worthy of note. The 1921–1922 peak includes the very earliest attempts to explain the origin of the isotope shift which had just been observed for the first time. The 1931–1932 peak shows the first application of

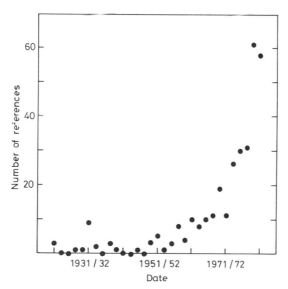

FIGURE 10.2. The distribution of references cited in this book with regard to their year of publication. Each point gives the number of references for a two-year period. The six references of prior to 1921 are omitted.

the then new quantum mechanics to the subject of isotope shifts. The 1971–1972 trough, if it is a genuine feature, has no obvious explanation; perhaps it reflects the period when experimental spectroscopists were starting to use lasers for the first time and were not able to make them work well enough to produce papers on isotope shifts in atomic spectra.

The subject of isotope shifts spans the fields of atomic and nuclear physics and, in the case of muonic atoms, particle physics as well. At the present time theory and experiment are well in step on the atomic side, but the theory lags behind on the nuclear side in the evaluation of nuclear charge distributions. The next big boost for the subject will probably be when nuclear theory can make good predictions of the nuclear charge distributions of individual nuclides. When this happens, the challenge will be to measure isotope shifts between isotopes that look particularly interesting from the theoretician's point of view, rather than between the isotopes that are most accessible to the wiles of the experimentalist.

SOME USEFUL QUANTITIES

In work on isotope shifts, some fundamental physical constants are sometimes needed with the ultimate accuracy. The following values are taken from British Standard number 5775. The figures in parenthesis represent the best estimate of the standard deviation uncertainties.

Speed of light in vacuum	$c = 2.997\ 924\ 580(12) \times 10^8\ \mathrm{ms}^{-1}$
Rest mass of electron	$m_e = 5.485\ 802\ 6(21) \times 10^{-4}\ \mathrm{u}$
Rest mass of proton	$m_p = 1.007\ 276\ 470(11)\ \mathrm{u}$
	$m_p/m_e = 1836.151\ 52(70)$
Rest mass of muon	$m_\mu = 0.113\ 429\ 20(26)\ \mathrm{u}$
Rydberg constant (fixed nucleus)	$R_\infty = 1.097\ 373\ 177(83) \times 10^7\ \mathrm{m}^{-1}$
	$= 3.289\ 842\ 00(25) \times 10^{15}\ \mathrm{Hz}$

Other constants that are not required to have such accuracy are given to six significant figures:

Permittivity of vacuum	$\varepsilon_0 = 8.85419 \times 10^{-12}\ \mathrm{F\ m}^{-1}$
Planck constant	$h = 6.62618 \times 10^{-34}\ \mathrm{J\ Hz}^{-1}$
	$h/2\pi = 6.58217 \times 10^{-16}\ \mathrm{eV\ s}$
Elementary charge	$e = 1.60219 \times 10^{-19}\ \mathrm{C}$
Atomic mass unit	$u = 1.66057 \times 10^{-27}\ \mathrm{kg}$
Bohr magneton	$\mu_B = 9.27408 \times 10^{-24}\ \mathrm{J\ T}^{-1}$
Nuclear magneton	$\mu_N = 5.05082 \times 10^{-27}\ \mathrm{J\ T}^{-1}$
First Bohr radius	$a_0 = 5.29177 \times 10^{-11}\ \mathrm{m}$
Fine structure constant	$\alpha = 1/137.036$
Rydberg constant	$R_\infty = 13.6058\ \mathrm{eV}$

Some non-SI units that frequently arise in the literature:

$$1\ \text{ångström } (\text{Å}) = 10^{-10}\ \mathrm{m}$$

$$1\ \text{millikayser } (\text{mK}) = 10^{-3}\ \mathrm{cm}^{-1} = 10\ \mathrm{m}^{-1}$$

Energies of photons can be expressed in various ways which are related as follows:

	Frequency	Wave number	Energy
1 MHz	1 MHz	3.3356×10^{-2} mK	4.1357×10^{-6} meV
1 mK	29.979 MHz	10^{-3} cm^{-1}	1.23985×10^{-4} meV
1 cm^{-1}	29.979 GHz	1000 mK	0.123985 meV
1 meV	241.80 GHz	8.0655 cm^{-1}	1 meV

Difference in frequency, etc.	Approximate wavelength difference at a wavelength of	
	500 nm = 5000 Å	30 pm = 300 mÅ
1 MHz	8.34×10^{-6} Å	3.00×10^{-11} mÅ
1 mK	2.50×10^{-4} Å	9.00×10^{-10} mÅ
1 cm^{-1}	0.25 Å	9.00×10^{-7} mÅ
1 meV	2.02 Å	7.26×10^{-6} mÅ

SELECTED REFERENCES

The following references are of more general interest than most of those in the main list of references. Details of these references may be found in the main list.

The subject of isotope shifts spans the fields of atomic and nuclear physics. An introductory text on atomic spectroscopy is that by Kuhn (1969) and a more detailed treatment is given by Cowan (1981). On the nuclear side, an introductory text is that by Segrè (1977) and nuclear structure is treated in great detail in the treatise by Bohr and Mottelson (1969; 1975; 1980). A book that concentrates on the interaction of atomic electrons with atomic nuclei is that by Kopfermann and Schneider (1958). Schneider's name is included in the reference since this author feels he should be credited with doing more than merely translating Kopfermann's book into English.

The first review article on isotope shifts was by Foster (1951) and this has been followed by several others such as those by Brix and Kopfermann (1958) and by Stacey (1966). The more recent reviews have concentrated on particular parts of the subject. Heilig and Steudel (1974) made an extremely thorough survey of the changes in the mean square radius of the nuclear charge distribution which could be obtained from optical isotope shifts. Bauche and Champeau (1976) reviewed the theory of isotope shifts in atomic spectra. Two bibliographies on experimental optical isotope shifts by Heilig (1977; 1982) are particularly useful, since papers in which "isotope shift" appears in neither the title nor the abstract are included. More general bibliographies covering the whole field of atomic energy levels and spectra, but indicating where isotope shifts are involved, are produced from time to time by the National Bureau of Standards, the latest being by Zalubas and Albright (1980).

There are, of course, other ways of finding out about nuclear charge distributions than by measuring isotope shifts and the subject is dealt with in a more general way by Barrett (1974) and by Barrett and Jackson (1977). The latter considers the distribution of nuclear matter as well as nuclear charge, and so does the more recent review of Hodgson (1981). Much of our knowledge of nuclear charge distributions comes from the study of the energies of muonic x-rays and this subject has recently been reviewed by Borie and Rinker (1982).

The measurement of isotope shifts in optical spectra requires very high resolution and suitable techniques are described in two volumes in the series

"Physics of Atoms and Molecules" edited by Hanle and Kleinpoppen (1978). Particularly relevant chapters are by Heilig and Steudel on new developments of classical optical spectroscopy and by Demtröder on laser high-resolution spectroscopy in Part A and by Kluge on optical spectroscopy of short-lived isotopes in Part B. High-resolution spectroscopy of radioactive atoms, particularly using laser techniques, is a subject that has also been reviewed by Jacquinot and Klapisch (1979) and by Otten (1981).

REFERENCES

Accad, Y., Pekeris, C. L., and Schiff, B., 1971, S and P states of the helium isoelectronic sequence up to $Z = 10$, *Phys. Rev. A* **4**, 516–536.

Ahmad, S. A., and Saksena, G. D., 1981, Isotope shifts in the energy levels of the singly ionized neodymium atom, *Spectrochim. Acta* **36B**, 943–950.

Ahmad, S. A., Venugopalan, A., and Saksena, G. D., 1979, Isotope shifts in odd and even energy levels of the neutral and singly ionized gadolinium atom, *Spectrochim. Acta* **34B**, 221–235.

Alvarez, E., Arnesen, A., Bengston, A., Hallin, R., Mayige, C., Nordling, C., Noreland, T., and Staaf, O., 1979, Isotope shift and hyperfine structure measurements in Xe II in a laser–ion beam experiment, *Phys. Scr.* **20**, 141–144.

Amin, S. R., Caldwell, C. D., and Lichten, W., 1981, Crossed-beam spectroscopy of hydrogen: a new value for the Rydberg constant, *Phys. Rev. Lett.* **47**, 1234–1238.

Andle, A., Bekk, K., Göring, S., Hanser, A., Nowicki, G., Rebel, H., Schatz, G., and Thompson, R. C., 1982, Isotope shifts and hyperfine structure of the $4s^2\,^1S_0 - 4s4p\,^1P_1$ transition in calcium isotopes, *Phys. Rev. C* **26**, 2194–2202.

Anigstein, R., Budick, B., and Kast, J. W., 1980, Neutron anomalous-moment contributions to muonic isomer shifts in ^{207}Pb, *Phys. Rev. Lett.* **44**, 1484–1487.

Aronberg, L., 1918, Note on the spectrum of the isotopes of lead, *Astrophys. J.* **47**, 96–101.

Aufmuth, P., 1978, Isotope shift and configuration mixing in dysprosium II, *Z. Phys. A* **286**, 235–241.

Aufmuth, P., 1982, Field shift crossed second-order effects in $l^N s$ configurations, *J. Phys. B* **15**, 3127–3140.

Aufmuth, P., Clieves H.-P., Heilig, K., Steudel, A., and Wendlandt, D., 1978, Isotope shift in molybdenum, *Z. Phys. A* **285**, 357–364.

Babushkin, F. A., 1963, Isotope shift of spectral lines, *Sov. Phys. JETP (USA)* **17**, 1118–1122.

Backenstoss, G., Kowald, W., Schwanner, I., Tauscher, L., and Weyer, H. J., 1980, Precision determination of the difference of the charge radii of ^{16}O and ^{18}O, *Phys. Lett.* **95B**, 212–214.

Baird, P. E. G., 1976, Isotope shifts and hyperfine structure in the atomic spectrum of palladium, *Proc. R. Soc. London, Ser. A* **351** 267–275.

Baird, P. E. G., Brambley, R. J., Burnett, K., Stacey, D. N., Warrington, D. M., and Woodgate, G. K., 1979, Optical isotope shifts and hyperfine structure in λ553.5 nm of barium, *Proc. R. Soc. London Ser. A* **365**, 567–582.

Baird, P. E. G., and Stacey, D. N., 1974, Isotope shifts and hyperfine structure in the optical spectrum of platinum, *Proc. R. Soc. London Ser. A* **341**, 399–406.

Ballik, E. A., 1972, Gain method for the measurement of isotope shift, *Can. J. Phys.* **50**, 47–51.

Barbier, L., and Champeau, R.-J., 1980, Very high resolution study of high Rydberg levels of the configurations $4f^{14}6snd$ of Yb I, *J. Phys. (France)* **41**, 947–955.

Bardin, T. T., Cohen, R. C., Devons, S., Hitlin, D., Macagno, E., Rainwater, J., Runge, K., Wu, C. S., and Barrett, R. C., 1967, Magnetic dipole and electric quadrupole hyperfine effects in Bi^{209} muonic x rays, *Phys. Rev.* **160**, 1043–1054.

Barreau, P., Roussel, L., and Powers, R. J., 1981, A muonic x-ray study of the charge distribution of ^{147}Sm and ^{149}Sm, *Nucl. Phys.* **A364**, 446–460.

Barrett, R. C., 1970, Model-independent parameters of the nuclear charge distribution from muonic x-rays, *Phys. Lett.* **33B**, 388–390.

Barrett, R. C., 1974, Nuclear charge distributions, *Rep. Prog. Phys.* **37**, 1–54.

Barrett, R. C., and Jackson, D. F., 1977, *Nuclear Sizes and Structure*, Oxford University Press, Oxford.

Barrett, R. C., Owen, D. A., Calmet, J., and Grotch, H., 1973, Recoil corrections to muonic atom energy levels, *Phys. Lett.* **47B**, 297–299.

Bartlett, J. H., 1931, Isotopic displacement in hyperfine structure, *Nature* **128**, 408–409.

Bauche, J., 1966, Évaluation théorique du déplacement isotopique spécifique, *C. R. Hebd. Seances Acad. Sci. Ser. B* **263**, 685–688.

Bauche, J., 1969, *Thèse*, Université de Paris.

Bauche, J., 1974, Hartree–Fock evaluations of specific-mass isotope shifts, *J. Phys. (France)* **35**, 19–26.

Bauche, J., 1981, On the use of $|\psi(0)|^2$ from hyperfine structure in isotope shift, *Comments Atom. Mol. Phys.* **10**, 57–68.

Bauche, J., and Champeau, R.-J., 1976, Recent progress in the theory of atomic isotope shift, *Adv. Atom. Mol. Phys.* **12**, 39–86.

Bauche, J., Champeau, R.-J., and Sallot, C., 1977, J-dependent isotope shifts in the ground term of samarium I, *J. Phys. B* **10**, 2049–2059.

Bauche, J., and Crubellier, A., 1970, Évaluation théorique des déplacements isotopiques spécifiques dans la série 3d, *J. Phys. (France)* **31**, 429–434.

Bayer, R., Kowalski, J., Neumann, R., Noehte, S., Suhr, H., Winkler, K., and zu Putlitz, G., 1979, Dye laser saturation spectroscopy of the $2^3S_1 - 2^3P$ transition in $^{6,7}Li^+$ ions, *Z. Phys. A* **292**, 329–336.

Beigang, R., and Timmermann, A., 1982, Level shifts in ms*ns* Rydberg states induced by hyperfine interaction in odd alkaline–earth isotopes, *Phys. Rev. A* **25**, 1496–1503.

Bekk, K., Andl, A., Göring, S., Hanser, A., Nowicki, G., Rebel, H., and Schatz, G., 1979, Laser spectroscopic studies of collective properties of neutron-deficient Ba nuclei, *Z. Phys. A* **291**, 219–230.

Bemis, C. E., Jr., Beene, J. R., Young, J. P., and Kramer, S. D., 1979, Optical isomer shift for the spontaneous-fission isomer $^{240}Am^m$, *Phys. Rev. Lett.* **43**, 1854–1858.

Bergmann, E., Bopp, P., Dorsch, Ch., Kowalski, J., Träger, F., and zu Putlitz, G., 1980, Nuclear charge distribution of eight Ca-nuclei by laser spectroscopy, *Z. Phys. A* **294**, 319–322.

Bernow, S., Devons, S., Deurdoth, I., Hitlin, D., Kast, J. W., Macagno, E. R., Rainwater, J., Runge, K., and Wu, C. S., 1967, Measurement of the nuclear gamma ray in muonic Sm^{152}, *Phys. Rev. Lett.* **18**, 787–791.

Bethe, H. A., and Salpeter, E. E., 1977, *Quantum Mechanics of One- and Two-Electron Atoms*, Plenum Press, New York.

Bhattacharjee, S. K., Boehm, F., and Lee, P. L., 1969, Nuclear charge radii from atomic K x rays, *Phys. Rev.* **188**, 1919–1929.

Biraben, F., Giacobino, E., and Grynberg, G., 1975, Doppler-free two-photon spectroscopy of neon, *Phys. Rev. A* **12**, 2444–2446.

Bi. aben, F., Grynberg, G., Giacobino, E., and Bauche, J., 1976, Investigation of isotopic shift in the 4d′ subconfiguration of neon using Doppler-free two-photon spectroscopy, *Phys. Lett.* **56A**, 441–442.

Bishop, D. C., and King, W. H., 1971, Isotope shifts in the resonance lines of Cd II and Sn IV and the size of specific mass shifts in alkali-like resonance lines, *J. Phys. B* **4**, 1798–1807.

Blaise, J., and Radziemski, Jr., L. J., 1976, Energy levels of neutral atomic uranium (U I), *J. Opt. Soc. Am.* **66**, 644–659.

Blaise, J., and Steudel, A., 1968, Isotopieverschiebungskonstanten von Th, U, Pu und Am, *Z. Phys.* **209**, 311–328.

Bodmer, A. R., 1953, Nuclear scattering of electrons and isotope shift, *Proc. Phys. Soc. A* **66**, 1041–1058.

Bodmer, A. R., 1959, Isotope shift and changes of nuclear radius, *Nucl. Phys.* **9**, 371–390.

Boehm, F., and Lee, P. L., 1974, Isotope shifts of electronic K_α X rays and variations of the mean square nuclear charge radii, *At. Data Nucl. Data Tables* **14**, 605–611.

Boerner, H., Harzer, R., Jitschin, W., Meisel, G., Matthews, H. G., and Penselin, S., 1978, Laser isotope enrichment using charge transfer ionization, *Opt. Commun.* **26**, 351–353.

Bohr, A., and Mottelson, B. R., 1969, *Nuclear Structure, Vol. I: Single-Particle Motion*, W. A. Benjamin, Reading, Massachusetts.

Bohr, A., and Mottelson, B. R., 1975, *Nuclear Structure, Vol. II: Nuclear Deformations*, W. A. Benjamin, Reading, Massachusetts.

Bohr, A., and Mottelson, B. R., 1980, *Nuclear Structure, Vol. III: Nucleonic Correlations*, The Benjamin/Cummings Publishing Company, Reading, Massachusetts.

Bohr, A., and Weisskopf, V. F., 1950, The influence of nuclear structure on the hyperfine structure of heavy elements, *Phys. Rev.* **77**, 94–98.

Bohr, N., 1913, On the constitution of atoms and molecules, *Philos. Mag.* **26**, 1–25.

Bohr, N., 1922, Remarks on the difference between series spectra of isotopes, *Nature* **109**, 746.

Bonn, J., Huber, G., Kluge, H.-J., and Otten, E. W., 1976, Spins, moments, and charge radii in the isotopic series ^{181}Hg–^{191}Hg*, *Z. Phys. A* **276**, 203–217.

Borghs, G., De Bisschop, P., Silverans, R. E., Van Hove, M., and Van den Cruyce, J. M., 1981, Hyperfine structures and isotope shifts of the $5d\,^4D_{7/2} \rightarrow 6p\,^4P_{5/2}{}^0$ transition in xenon ions, *Z. Phys. A* **299**, 11–13.

Borie, E., and Rinker, G. A., 1982, The energy levels of muonic atoms, *Rev. Mod. Phys.* **54**, 67–118.

Bradley, L. C., and Kuhn, H., 1951, Isotope shifts in the spectrum of helium, *Proc. R. Soc. London Ser. A* **209**, 325–333.

Brand, H., Pfeufer, V., and Steudel, A., 1981, Laser–atomic beam spectroscopy of $4f^75d6s$–$4f^75d6p$ transitions in Eu I, *Z. Phys. A* **302**, 291–298.

Brand, H., Seibert, B., and Steudel, A., 1980, Laser–atomic beam spectroscopy in Sm: Isotope shifts and changes in mean-square nuclear charge radii, *Z. Phys. A* **296**, 281–286.

Brandt, H.-W., Heilig, K., Knöckel, H., and Steudel, A., 1978, Isotope shift in the Ca I resonance line and changes in mean-square nuclear charge radii of the stable Ca isotopes, *Z. Phys. A* **288**, 241–246.

Brault, J. W., 1982, Fourier transform spectrometry in relation to other passive spectrometers, *Phil. Trans. R. Soc. London Ser. A* **307**, 503–511.

Bréchignac, C., 1977, Measurements of isotope shift in visible lines of Kr I by saturated-absorption techniques, *J. Phys. B* **10**, 2105–2110.

Brechignac, C., Gerstenkorn, S., and Luc, P., 1976, Déplacement isotopique dans la raie de resonance D$_2$ du rubidium. Inversion des valeurs de $\langle r^2 \rangle$ des isotopes ^{85}Rb et ^{87}Rb, *Physica* **82C**, 409–417.

Breit, G., 1932, The isotope displacement in hyperfine structure, *Phys. Rev.* **42**, 348–354.

Brimicombe, M. S. W. M., Stacey, D. N., Stacey, V., Hühnermann, H., and Menzel, N., 1976, Optical isotope shifts and hyperfine structure in Cd, *Proc. R. Soc. London, Ser. A* **352**, 141–152.

Briscoe, W. J., Crannell, H., and Bergstrom, J. C., 1980, Elastic electron scattering from the isotopes ^{35}Cl and ^{37}Cl, *Nucl. Phys. A* **344**, 475–488.

Brix, P., and Kopfermann, H., 1949, Zur Isotopieverschiebung im Spektrum des Samariums, *Z. Phys.* **126**, 344–364.

Brix, P., and Kopfermann, H., 1952, "Hyperfeinstruktur der Atomkerne und Atomlinien" in *Landolt-Börnstein, Zahlenwerte und Funktionen*, Vol. 1, Part 5, Springer-Verlag, Berlin.

Brix, P., and Kopfermann, H., 1958, Isotope shift studies of nuclei, *Rev. Mod. Phys.* **30**, 517–520.

Broadhurst, J. H., Cage, M. E., Clark, D. L., Greenlees, G. W., Griffith, J. A. R., and Isaak, G. R., 1974, High-resolution measurements of isotope shifts and hyperfine splittings for ytterbium using a cw tunable laser, *J. Phys. B* **7**, L513–L517.

Broch, E. K., 1945, On the evaluation of the isotope shift in hyperfine structure, *Arch. Math. Naturvidenskab B* **48**, 25–35.

Brochard, J., Cahuzac, P., and Vetter, R., 1967, Mesure des écarts isotopiques de six raies laser infrarouges dans l'argon, *C. R. Hebd. Seances Acad. Sci. Ser. B* **265**, 467–470.

Brochard, J., and Vetter, R., 1966, Méthode de mesure de faibles écarts isotopiques dans les émissions stimulées infrarouges, *C. R. Hebd. Seances Acad. Sci. Ser. B* **262**, 681–684.

Brockmeier, R. T., Boehm, F., and Hatch, E. N., 1965, Observation of the isotope shift of K_{α_1} X rays of uranium, *Phys. Rev. Lett.* **15**, 132–135.

Brown, B. A., Massen, S. E., and Hodgson, P. E., 1979, Proton and neutron density distributions for $A = 16$–58 nuclei, *J. Phys. G* **5**, 1655–1698.

Bruch, R., Heilig, K., Kaletta, D., Steudel, A., and Wendlandt, D., 1969, Nuclear volume and mass effect in the optical isotope shift of light elements, *J. Phys. (France)* **30**, C1.51–C1.58.

Buchholz, B., Kronfeldt, H.-D., Müller, G., Voss, M., and Winkler, R., 1978, Electric and magnetic hyperfine structure investigations in the $5s^2 5p^3$ and $5s^2 5p^2 6s$ configurations of ^{121,123}Sb, *Z. Phys. A* **288**, 247–246.

Buchholz, B., Kronfeldt, H.-D., and Winkler, R., 1979, Level classification in the configuration $5d^4 6s^2 6p$ of ^{185,187}Re I by isotope shift measurements, *Physica* **96C**, 297–301.

Bürger, K. H., Burghardt, B., Büttgenbach, S., Harzer, R., Hoeffgen, H., Meisel, G., and Träber, F., 1982, Hyperfine structure and isotope shift of high-lying metastable states of rhenium, *Z. Phys. A* **307**, 201–209.

Burghardt, B., Harzer, R., Meisl, G., and Penselin, S., 1980, Isotope separation of Dy atoms using laser excitation followed by charge exchange with Cs^+ ions, *Opt. Commun.* **33**, 169–172.

Burke, E. W., Jr., 1955, Isotope shift in the first three spectra of boron, *Phys. Rev.* **99**, 1839–1841.

Cagnac, B., 1982, High-resolution spectroscopy with multiple-beam laser techniques, *Phil. Trans. R. Soc. London, Ser. A* **307**, 633–644.

Čajko, J., 1970, Isotope shifts in Hf I spectrum and deformation of even nuclei of hafnium isotopes, *Z. Phys.* **234**, 443–454.

Campi, X., and Epherre, M., 1980, Calculation of binding energies and isotope shifts in a long series of rubidium isotopes, *Phys. Rev. C* **22**, 2605–2608.

Cavedon, J., Bellicard, J. B., Frois, B., Goutte, D., Huet, M., Leconte, P., Phan, X.-H., and Platchkov, S. K., 1982, The charge density distributions of ^{116}Sn and ^{124}Sn, *Phys. Lett.* **118B**, 311–314.

Champeau, R.-J., 1972, Déplacement isotopique dans les spectre d'arc et d'étincelle du cérium, *Physica* **62**, 209–238.

Champeau, R.-J., and Keller, J.-C., 1978, Spectroscopie laser à très haute résolution sur un jet atomique de krypton, *J. Phys. B* **11**, 391–397.

Champeau, R.-J., Michel, J.-J., and Walther, H., 1974, Spectroscopic determination of the nuclear moments of ^{169}Yb; relative isotope shift between the isotopes ^{168}Yb, ^{169}Yb and ^{170}Yb, *J. Phys. B* **7**, L262–L265.

Champeau, R.-J., and Miladi, M., 1974, Déplacement isotopique relatif dans le spectre d'arc du tungstène. Structure hyperfine du niveau $5d^5 6s^7 S_3$ de W I, *J. Phys. (France)* **35**, 105–111.

Champeau, R.-J., and Verges, J., 1976, Déplacement isotopique dans le spectre infrarouge du cérium, *Physica* **83C**, 373–378.

Charles, G. W., 1966, Spectra of ^{208}Po and the hyperfine structure of ^{209}Po, *J. Opt. Soc. Am.* **56**, 1292–1297.

Chesler, R. B., and Boehm, F., 1968, Isotope shifts of K X rays and variations of nuclear charge radii, *Phys. Rev.* **166**, 1206–1212.

Childs, W. J., and Goodman, L. S., 1979, ^{138,139}La nuclear electric–quadrupole-moment ratio by laser–rf double resonance, *Phys. Rev. A* **20**, 1922–1926.

Chuckrow, R., and Stroke, H. H., 1971, Hyperfine structure and isotope shift of 30-year ^{207}Bi in the 3067-Å resonance line, *J. Opt. Soc. Am.* **61**, 218–222.

Clark, D. L., Cage, M. E., Lewis, D. A., and Greenlees, G. W., 1979, Optical isotopic shifts and hyperfine splittings for Yb, *Phys. Rev. A* **20**, 239–253.

Close, D. A., Malanify, J. J., and Davidson, J. P., 1978, Nuclear charge distributions deduced from the muonic atoms of ^{232}Th, ^{235}U, ^{238}U, and ^{239}Pu, *Phys. Rev. C* **17**, 1433–1455.

Cohen, R. C., Devons, S., Kanaris, A. D., and Nissim-Sabat, C., 1966, Isotope shift in the muonic K X rays of O^{16}–O^{18}, Ca^{40}–Ca^{44}, Sn^{116}–Sn^{124}, and $Gd^{155-160}$, *Phys. Rev.* **141**, 48–55.

Coillaud, B., Bloomfield, L. A., Lawler, J. E., Siegel, A., and Hänsch, T. W., 1980, Saturation spectroscopy of ultraviolet transitions in mercury with a frequency-doubled cw ring dye laser, *Opt. Commun.* **35**, 359–362.

Coulthard, M. A., 1973, Relativistic contributions to screening ratios in europium, *J. Phys. B* **6**, 23–29.

Cowan, R. D., 1981, *The Theory of Atomic Structure and Spectra*, University of California Press, Berkeley.

Cowan, R. D., and Griffin, D. C., 1976, Approximate relativistic corrections to atomic radial wavefunctions, *J. Opt. Soc. Am.* **66**, 1010–1014.

Crawford, M. F., and Schawlow, A. L., 1949, Electron–nuclear potential fields from hyperfine structure, *Phys. Rev.* **76**, 1310–1317.

Dabkiewicz, P., Buchinger, F., Fischer, H., Kluge, H. J., Kremmling, H., Kühl, T., Müller, A. C., and Schuessler, H. A., 1979, Nuclear shape isomerism in ^{185}Hg detected by laser spectroscopy, *Phys. Lett.* **82B**, 199–203.

de Clercq, E., Biraben, F., Giacobino, E., Grynberg, G., and Bauche, J., 1981, Isotopic shift in helium, *J. Phys. B* **14**, L183–L186.

Demtröder, W., 1978, in: *Progress in Atomic Spectroscopy* (W. Hanle and H. Kleinpoppen, eds.) Part A, pp. 679–712, Plenum Press, New York.

Ducas, T. W., Littman, M. G., Freeman, R. R., and Kleppner, D., 1975, Stark ionization of high-lying states of sodium, *Phys. Rev. Lett.* **35**, 366–369.

Duke, C., Fischer, H., Kluge, H.-J., Kremmling, H., Kühl, Th., and Otten, E.-W., 1977, Determination of the isotope shift of ^{190}Hg by on-line laser spectroscopy, *Phys. Lett.* **60A**, 303–306.

Edlén, B., Ölme, A., Herzberg, G., and Johns, J. W. C., 1970, Ionization potential of boron, and the isotopic and fine structure of $2s2p^{2}$ ^{2}D, *J. Opt. Soc. Am.* **60**, 889–891.

Edwin, R. P., and King, W. H., 1969, Isotope shift measurements in the spectrum of trebly ionized cerium (Ce IV), *J. Phys. B* **2**, 260–267.

Ehrenfest, P., 1922, The difference between series spectra of isotopes, *Nature* **109**, 745–746.

Ehrlich, R. D., 1968, Precise muonic K_α isotope shifts for nuclides from Si to Sn, *Phys. Rev.* **173**, 1088–1100.

Eichhorn, A., Elbel, M., Kamka, W., Quad, R., and Bauche, J., 1982, Laser spectroscopic measurement of isotope shifts of transitions 3d–4p in the ionic spectra of argon, chlorine and sulphur, *Z. Phys. A* **305**, 39–44.

Elbel, M., Fischer, W., and Hartmann, M., 1963, Hyperfeinstruktur und Isotopieverschiebung im Kupfer II-Spektrum, *Z. Phys.* **176**, 288–292.

Elbel, M., and Hühnermann, H., 1969, Exchange polarization of the 3d-shell and isotope shift of the levels Cu II $3d^{9}4s$ ^{1}D and ^{3}D, *J. Phys. (France)* **30**, C1.41–C1.47.

Eliel, E. R., Hogervorst, W., Van Leeuwen, K. A. H., and Post, B. H., 1981, A frequency-doubled frequency-stabilized cw ring dye laser for spectroscopy. A study of the $\lambda = 293.3$ nm transition in In I, *Opt. Commun.* **36**, 366–368.

Engfer, R., Schneuwly, H., Vuillemier, J. L., Walter, H. K., and Zehnder, A., 1974, Charge-distribution parameters, isotope shifts, isomer shifts, and magnetic hyperfine constants from muonic atoms, *At. Data & Nucl. Data Tables* **14**, 509–597.

Engleman, R., Jr., Keller, R. A., and Palmer, B. A., 1980, Hyperfine structure and isotope shift of the 1.3-μm transition of ^{129}I, *Appl. Opt.* **19**, 2767–2770.

Engleman, R., Jr., and Palmer, B. A., 1980, Precision isotope shifts for the heavy elements. 1. Neutral uranium in the visible and near infrared. *J. Opt. Soc. Am.* **70**, 308–317.

Erickson, G. W., 1977, Energy levels of one-electron atoms, *J. Phys. & Chem. Ref. Data* **6**, 831–869.

Euteneuer, H., Friedrich, J., and Voegler, N., 1978, The charge-distribution differences of ^{209}Bi, ^{208,207,206,204}Pb and ^{205,203}Tl investigated by elastic electron scattering and muonic x-ray data, *Nucl. Phys.* **A298**, 452–476.

Fajardo, L. A., Ficenec, J. R., Trower, W. P., and Sick, I., 1971, Elastic electron–zirconium scattering, *Phys. Lett.* **37B**, 363–365.

Ferguson, A. I., 1982, Optoacoustic and optogalvanic spectroscopy, *Phil. Trans. R. Soc. London*, *Ser. A* **307**, 645–657.

Fermi, E., and Segrè, E., 1933, Zur Theorie der Hyperfeinstruktur, *Z. Phys.* **82**, 729–749.

Fischer, C. F., 1970, A multi-configuration Hartree–Fock program, *Comput. Phys. Commun.* **1**, 151–166.

Fischer, W., Hartmann, M., Hühnermann, H., and Vogg, H., 1974a, Isotopieverschiebung der Bariumnuklide ^{140}Ba, ^{138}Ba, ^{136}Ba und ^{134}Ba in der Ba II-Resonanzlinie $6p\,^2P_{1/2}-6s\,^2S_{1/2}$, *Z. Phys.* **267**, 209–217.

Fischer, W., Hühnermann, H., Krömer, G., and Schäfer, H. J., 1974b, Isotope shifts in the atomic spectrum of xenon and nuclear deformation effects, *Z. Phys.* **270**, 113–120.

Fischer, W., Hühnermann, H., and Mandrek, K., 1974c, Isotope shift measurements in the atomic spectrum of lanthanum (La I), *Z. Phys. A* **269**, 245–252.

Fischer, W., Hühnermann, H., Mandrek, K., Meier, Th., and Aumann, D. C., 1975a, Optical isotope shift in ^{144}Ce, *Physica* **79C**, 105–112.

Fischer, W., Hühnermann, H., and Meier, Th., 1975b, Nuclear moments and optical isotope shifts of ^{108m}Ag and ^{110m}Ag, *Z. Phys. A* **274**, 79–85.

Flusberg, A., Mossberg, T., and Hartmann, S. R., 1976, Hyperfine structure, isotopic level shifts, and pressure self-broadening of the $7\,^2P$ states of natural thallium by Doppler-free two-photon absorption, *Phys. Rev. A* **14**, 2146–2158.

Foldy, L. L., 1958, Fermi–Segrè formula, *Phys. Rev.* **111**, 1093–1098.

Foot, C. J., Stacey, D. N., Stacey, V., Kloch, R., and Leś, Z., 1982, Isotope effects in the nuclear charge distribution in zinc, *Proc. R. Soc. London, Ser. A* **384**, 205–216.

Ford, K. W., and Rinker, G. A., 1973, Analysis of muonic-atom X rays in the lead isotopes, *Phys. Rev. C* **7**, 1206–1221.

Foster, E. W., 1951, Nuclear effects in atomic spectra, *Rep. Prog. Phys.* **14**, 288–315.

Fradkin, E. E., 1962, Isotope shift of spectral lines and the compressibility of deformed nuclei, *Sov. Phys. JETP (USA)* **15**, 550–557.

Fred, M., and Tomkins, F. S., 1957, Preliminary term analysis of Am I and Am II spectra, *J. Opt. Soc. Am.* **47**, 1076–1087.

Freeman, R. R., Liao, P. F., Panock, R., and Humphrey, L. M., 1980, Isotope shift of the $2\,^3P-3\,^3D$ transition in helium, *Phys. Rev. A* **22**, 1510–1516.

Friar, J. L., and Negele, J. W., 1975, Theoretical and experimental determination of nuclear charge distributions, *Adv. Nucl. Phys.* **8**, 219–376.

Gagné, J.-M., Nguyen, Van, S., Saint-Dizier, J.-P., and Pianarosa, P., 1976, Isotope shift of ^{234}U, ^{236}U, ^{238}U in U I, *J. Opt. Soc. Am.* **66**, 1415–1416.

Gagné, J.-M., Saint-Dizier, J.-P., and Pianarosa, P., 1977, Odd–even staggering of ^{235}U from the 5027 Å line in U I, *Opt. Commun.* **20**, 269–270.

Gagné, J.-M., Saint-Dizier, J.-P., and Pianarosa, P., 1978, Isotope shift ^{238}U–^{233}U from some lines in the U I spectrum, *Opt. Commun.* **26**, 348–350.

Garton, W. R. S., Reeves, E. M., and Tomkins, F. S., 1974, Hyperfine structure and isotope shift of the $6s6p^2\,^4P_{1/2}$ level of Tl I, *Proc. R. Soc. London, Ser. A* **341**, 163–166.

Gerhardt, H., Matthias, E., Rinneberg, H., Schneider, F., Timmermann, A., Wenz, R., and West, P. J., 1979, Changes in nuclear mean square charge radii of stable krypton isotopes, *Z. Phys. A* **292**, 7–14.

Gerhardt, H., Matthias, E., Schneider, F., and Timmermann, A., 1978, Isotope shifts and hyperfine structure of the 6s–7p transitions in the cesium isotopes 133, 135, and 137, *Z. Phys. A* **288**, 327–333.

Gerstenkorn, S., 1971, Relation entre énergies de liaison et déplacement isotopique relatif dans les raies spectrales des éléments possédant un nombre de protons voisin de 50 (cadmium, étain, tellure), *C. R. Hebd. Seances Acad. Sci. Ser. B* **272**, 110–113.

Gerstenkorn, S., 1973, Propriétés nucléaires déduites des spectres optique des atomes. Moments nucléaires et déplacements isotopique dans les actinides et dans la série des terres rares, *J. Phys. (France)* **34**, C4/55–C4/62.

Gerstenkorn, S., Labarthe, J. J., and Vergès, J., 1977, Fine and hyperfine structures and isotope shifts in the arc spectrum of mercury, *Phys. Scr.* **15**, 167–176.

Gerstenkorn, S., and Vergès, J., 1975, Interprétation des déplacements isotopiques pairs–impairs anormaux dans le spectre d'arc du mercure, *J. Phys. (France)* **36**, 481–486.

Giacobino, E., Biraben, F., de Clercq, E., Wohrer-Beroff, K., and Grynberg, G., 1979, Doppler-free two-photon spectroscopy of neon. III. Isotopic shift and fine structure for the $2p^5 4d$ and $2p^5 5s$ configurations, *J. Phys. (France)* **40**, 1139–1144.

Gillespie, W. A., Brain, S. W., Johnston, A., Lees, E. W., Singhal, R. P., Slight A. G., Brimicombe, M. W. S. M., Stacey, D. N., Stacey, V., and Hühnermann H., 1975, Measurements of the nuclear charge distribution in the cadmium isotopes, *J. Phys. G* **1**, L6–L8.

Gluck, G. G., Bordarier, Y., Bauche, J., and Van Kleef, T. A. M., 1964, Isotopic displacement, theoretical calculations, and structure of the terms in the arc spectrum of osmium, *Physics* **30**, 2068–2104.

Goble, A. T., Silver, J. D., and Stacey, D. N., 1974, Isotope shifts in the atomic spectrum of tin: ^{112}Sn, ^{114}Sn and ^{115}Sn, *J. Phys. B* **7**, 26–30.

Goorvitch, D., Davis, S. P., and Kleiman, H., 1969, Isotope shift and hyperfine structure of the neutron-deficient thallium isotopes, *Phys. Rev.* **188**, 1897–1904.

Goudsmit, S., 1933, Nuclear magnetic moments, *Phys. Rev.* **43**, 636–639.

Gräff, G., Kalinowsky, H., and Traut, J., 1980, A direct determination of the proton electron mass ratio, *Z. Phys. A* **297**, 35–39.

Grafström, P., Zhan-Kui, J., Jönsson, G., Kröll, S., Levinson, C., Lundberg, H., and Svanberg, S., 1982, Hyperfine structure and isotope shift of highly excited barium-I states, *Z. Phys. A* **306**, 281–284.

Grant, I. P., 1970, Relativistic calculation of atomic structures, *Adv. Phys.* **19**, 747–811.

Grant, I. P., 1980, Many-electron effects in the theory of nuclear volume isotope shift, *Phys. Scr.* **21**, 443–447.

Grethen, H., Winkler, R., and Bauche, J., 1980, Experimental investigations and parametric analysis of the isotope shift of the levels of the ground configurations $5d^8 6s^2$ and $5d^9 6s$ in the Pt I spectrum, *Physica* **98C**, 222–228.

Griffith, J. A. R., 1982, Laser heterodyne spectroscopy, *Phil. Trans. R. Soc. London, Ser. A* **307**, 563–571.

Griffith, J. A. R., Isaak, G. R., New, R., and Ralls, M. P., 1981, Anomalies in the optical isotope shifts of samarium, *J. Phys. B* **14**, 2769–2780.

Griffith, J. A. R., Isaak, G. R., New, R., Ralls, M. P., and van Zyl, C. P., 1977, Optical heterodyne spectroscopy using tunable dye lasers: hyperfine structure of sodium, *J. Phys. B* **10**, L91–L95.

Griffith, J. A. R., Isaak, G. R., New, R., Ralls, M. P., and van Zyl, C. P., 1979, Anomalies in the optical isotope shifts of samarium, *J. Phys. B* **12**, L1–L7.

Grundevik, P., Gustavsson, M., Rosen, A., and Rydberg, S., 1979, Analysis of the isotope shifts and hyperfine strucutre in the 3988 Å (6s6p^{1}P$_1$ ↔ 6s^2 ^{1}S$_0$) Yb I Line, *Z. Phys. A* **292**, 307–310.

Grundevik, P., Lundberg, H., Nilsson, L., and Olsson, G., 1982, Hyperfine structure in the 6p5d configuration and isotope shifts in transitions between the 6s5d and the 6p5d configurations in Ba I, *Z. Phys. A* **306**, 195–209.

Güttinger, P., and Pauli, W., 1931, Zur hyperfeinstruktur von Li$^+$, *Z. Phys.* **67**, 743–765.

Hahn, A. A., Miller, J. P., Powers, R. J., Zehnder, A., Rushton, A. M., Welsh, R. E., Kunselman, A. R., Roberson, P., and Walter, H. K., 1979, An experimental study of muonic x-ray transitions in mercury isotopes, *Nucl. Phys.* **A314**, 361–386.

Hallstadius, L., 1979, Extended measurements of isotope shifts in Mg I, *Z. Phys. A* **291**, 203–206.

Hanle, W., and Kleinpoppen, H., (eds.), 1978, *Progress in Atomic Spectroscopy* (2 vols), Plenum Press, New York.

Hansch, T. W., Lyons, D. R., Schawlow, A. L., Siegel, A., Wang, Z.-Y., and Yan, G.-Y., 1981, Polarization intermodulated excitation (Polinex) spectroscopy of helium and neon, *Opt. Commun.* **38**, 47–51.

Hansch, T. W., Nayfeh, M. H., Lee, S. A., Curry, S. M., and Shahin, I. S., 1974, Precision measurement of the Rydberg constant by laser saturation spectroscopy of the Balmer α line in hydrogen and deuterium, *Phys. Rev. Lett.* **32**, 1336–1340.

Hansen, J. E., Steudel, A., and Walther, H., 1967, Isotopieverschiebung der natürlichen geraden und ungeraden Sm- und Nd-isotope, *Z. Phys.* **203**, 296–329.

Hartley, H., Ponder, A. O., Bowen, E. W., and Merton, T. R., 1922, An attempt to separate the isotopes of chlorine, *Philos. Mag.* **43**, 430–435.

Heilig, K., 1961, Die Isotopieverschiebung zwischen den geraden Sr-Isotopen 84, 86, 88 und 90 und der Sprung im Kernvolumeneffekt bei der Neutronenzahl 50, *Z. Phys.* **161**, 252–266.

Heilig, K., 1977, Bibliography on experimental optical isotope shifts 1918 through October 1976, *Spectrochim. Acta* **32B**, 1–59.

Heilig, K., 1982, Bibliography on experimental optical isotope shifts: Part II, November 1976 to October 1981, *Spectrochim. Acta* **37B**, 417–455.

Heilig, K., Riesner, D., and Steudel, A., 1966, Isotope shift in Ge I, *J. Opt. Soc. Am.* **56**, 1406–1407.

Heilig, K., Schmitz, K., and Steudel, A., 1963, Isotopieverschiebung im Zirkon I-Spektrum, *Z. Phys.* **176**, 120–125.

Heilig, K., and Steudel, A., 1974, Changes in mean-square nuclear charge radii from optical isotope shifts, *At Data & Nucl. Data Tables* **14**, 613–638.

Heilig, K., and Steudel, A., 1978, in: *Progress in Atomic Spectroscopy* (W. Hanle and H. Kleinpoppen, eds.), Part A, pp. 263–328, Plenum Press, New York.

Heilig, K., and Wendlandt, D., 1967, Isotope shift in Cr I, *Phys. Lett.* **25A**, 277–278.

Herzberg, G., 1958, Ionization potentials and Lamb shifts of the ground states of ^{4}He and ^{3}He, *Proc. R. Soc. London, Ser. A* **248**, 309–332.

Hines, A. P., and Ross, J. S., 1962, Isotope shift in the spectrum of osmium, *Phys. Rev.* **126**, 2105–2108.

Hitlin, D., Bernow, S., Devons, S., Duerdoth, I., Kast, J. W., Macagno, E. R., Rainwater, J., Wu, C. S., and Barrett, R. C., 1970, Muonic atoms I. Dynamic hyperfine structure in the spectra of deformed nuclei, *Phys. Rev. C* **1**, 1184–1201.

Hodgson, P. E., 1981, Nuclear charge and matter distributions, *Contemp. Phys.* **22**, 511–532.

Hoehn, M. V., and Shera, E. B., 1979, Muonic resonance excitation of ^{188}Os and ^{172}Yb, *Phys. Rev. C* **20**, 1934–1941.

Hoehn, M. V., Shera, E. B., and Steffen, R. M., 1982, Muonic isomer shifts of the 803- and 2648-keV states, in ^{206}Pb, *Phys. Rev. C* **26**, 2242–2246.

Hoehn, M. V., Shera, E. B., and Wohlfahrt, H. D., 1980, Muonic isomer shift of the first excited 2$^+$ state in ^{204}Pb, *Phys. Rev. C* **22**, 678–680.

Hoehn, M. V., Shera, E. B., Wohlfahrt, H. D., Yamazaki, Y., Steffen, R. M., and Sheline, R. K., 1981, Muonic x-ray study of the even Os nuclei, *Phys. Rev. C* 24, 1667–1690.

Höhle, C., Hühnermann, H., Meier, Th., and Wagner, H., 1978, High-resolution spectroscopy of the transition $5d\,^2D_{3/2} \rightarrow 6p\,^2P^0_{3/2}$ in a fast Ba^+ ion beam, *Z. Phys. A* 284, 261–265.

Huber, G., Bonn, J., Kluge, H.-J., and Otten, E. W., 1976, Nuclear radiation detected optical pumping of neutron-deficient Hg isotopes, *Z. Phys. A* 276, 187–202.

Huber, G., Touchard, F., Büttgenbach, S., Thibault, C., Klapisch, R., Duong, H. T., Liberman, S., Pinard, J., Vialle, J. L., Juncar, P. and Jacquinot, P., 1978, Spins, magnetic moments, and isotope shifts of $^{21-31}$Na by high-resolution laser spectroscopy of the atomic D_1 line, *Phys. Rev. C* 18, 2342–2354.

Hughes, D. S., and Eckart, C., 1930, The effect of the motion of the nucleus on the spectra of Li I and Li II, *Phys. Rev.* 36, 694–698.

Hughes, R. H., 1955, Isotope shift in the first spectrum of atomic lithium, *Phys. Rev.* 99, 1837–1839.

Hühnermann, H., Valentin, H., and Wagner, H., 1978, Nuclear moments and optical isotope shift of ^{133}Xe, *Z. Phys. A* 285, 229–230.

Hühnermann, H., and Wagner, H., 1967, Isotope shift in the resonance lines of the arc spectrum of ^{133}Cs, ^{135}Cs, and ^{137}Cs, *Z. Phys.* 199, 239–243.

Hull, R. J., and Stroke, H. H., 1961, Nuclear moments and isotope shifts of Tl^{199}, Tl^{200}, Tl^{201}, Tl^{202}, and Tl^{204} —Isotope shifts in odd–odd nuclei. *Phys. Rev.* 122, 1574–1575.

Iwinski, Z. R., Kim, Y. S., and Pratt, R. H., 1980, Generalized Fermi–Segrè formula, *Phys. Rev. A* 22, 1358–1360.

Jackson, D. A., 1932, Die Hyperfeinstruktur der Thallium–Bogenlinien, *Z. Phys.* 75, 223–228.

Jackson, D. A., 1961, The spherical Fabry–Perot interferometer as an instrument of high resolving power for use with external or with internal atomic beams, *Proc. R. Soc. London, Ser. A* 263, 289–308.

Jackson, D. A., 1980, Isotope shifts in visible lines of the arc spectrum of krypton, *J. Opt. Soc. Am.* 70, 1139–1144.

Jackson, D. A., 1981, Hyperfine structure and isotope shifts in the lines 4033 Å of gallium I and 4101 Å of indium I, *Physica* 103C, 437–438.

Jackson, D. A., Coulombe, M.-C., and Bauche, J., 1975, Interpretation of the isotope shifts in the arc spectrum of xenon, *Proc. R. Soc. London, Ser. A* 343, 443–451.

Jackson, D. A., and Kuhn, H., 1938, Hyperfine structure, Zeeman effect, and isotope shift in the resonance lines of potassium, *Proc. R. Soc. London, Ser. A* 165, 303–312.

Jacquinot, P., and Klapisch, R., 1979, Hyperfine spectroscopy of radioactive atoms, *Rep. Prog. Phys.* 42, 773–832.

Jänecke, J., 1981, Simple parameterization of nuclear deformation parameters, *Phys. Lett.* 103B, 1–4.

Jitschin, W., and Meisel, G., 1980, Doppler-free two-photon laser spectroscopy of barium I: Hyperfine splitting and isotope shift of high-lying levels, *Z. Phys. A* 295, 37–43.

Julien, L., Pinard, M., and Laloë, F., 1980, Hyperfine structure and isotope shift of the 640.2 and 626.6 nm lines of neon, *J. Phys. (France)* 41, L479–L482.

Kasper, J. V. V., Pollock, C. R., Curl, R. F., Jr., and Tittel, F. K., 1981, Observation of the $^2P_{1/2} \leftarrow\ ^2P_{3/2}$ transition of the Br atom by color center laser spectroscopy, *Chem. Phys. Lett.* 77, 211–213.

Kast, J. W., Bernow, S., Cheng, S. C., Hitlin, D., Lee, W. Y., Macagno, E. R., Rushton, A. M., and Wu, C. S., 1971, The isotone shift in muonic X rays in the tin region, *Nucl. Phys.* A169, 62–70.

Kaufman, S. L., 1976, High-resolution laser spectroscopy in fast beams, *Opt. Commun.* 17, 309–312.

Keller, J. C., 1973, Parametric study of isotope shifts in Ne I, *J. Phys. B* 6, 1771–1778.

Kelly, F. M., 1957, Isotope shift in the resonance lines of Mg I, *Can. J. Phys.* 35, 1220–1222.

Kessler, D., Mes, H., Thompson, A. C., Anderson, H. L., Dixit, M. S., Hargrove, C. K., and McKee, R. J., 1975, Muonic X rays in lead isotopes, *Phys. Rev. C* **11**, 1719–1734.

King, W. H., 1963, Comments on the article "Peculiarities of the isotope shift in the samarium spectrum", *J. Opt. Soc. Am.* **53**, 638–639.

King, W. H., 1964, Isotope shift in the arc spectrum of ruthenium, *Proc. R. Soc. London, Ser. A* **280**, 430–438.

King, W. H., 1979, Isotope shift and configuration interaction in U I, *J. Phys. B* **12**, 383–386.

King, W. H., 1981, A suggestion that the anomalous isotope shifts in samarium are predominantly mass shifts, *J. Phys. B* **14**, L721–L723.

King, W. H., Kuhn, H. G., and Stacey, D. N., 1966, Isotope shifts in medium-heavy elements, *Proc. R. Soc. London, Ser. A* **296**, 24–37.

King, W. H., Steudel, A., and Wilson, M., 1973, Optical isotope shifts in neodymium, *Z. Phys.* **265**, 207–224.

King, W. H., and Wilson, M., 1971, Screening effects in optical isotope shifts in Ce IV, Sn IV and Cd II, *Phys. Lett.* **37A**, 109–110.

Kleiman, H., and Davis, S. P., 1963, Hyperfine structures, isotope shifts, and nuclear moments of Hg^{195}, Hg^{195m}, and Hg^{194}, *J. Opt. Soc. Am.* **53**, 822–827.

Klempt, W., Bonn, J., and Neugart, R., 1979, Nuclear moments and charge radii of neutron-rich Rb isotopes by fast-beam laser spectroscopy, *Phys. Lett.* **82B**, 47–50.

Kluge, H.-J., 1978, in: *Progress in Atomic Spectroscopy* (W. Hanle and H. Kleinpoppen, eds.), Part B, pp. 713–768, Plenum Press, New York.

Kluge, H.-J., Kremmling, H., Schuessler, H. A., Streib, J., and Wallmeroth, K., 1983, Determination of the isotope shift in the D_1 line between ^{197}Au and ^{195}Au, *Z. Phys. A* **309**, 187–192.

Kopfermann, H., and Krüger, H., 1937, Über die Anreicherung des Argonisotops Ar^{36} und den Isotopieverschiebungseffekt im Spektrum des Ar I, *Z. Phys.* **105**, 389–394.

Kopfermann, H., and Schneider, E. E., 1958, *Nuclear Moments*, Academic Press, New York.

Kotlikov, E. N., and Tokarev, V. I., 1980, Determination of the isotopic shift of the 632.8 nm neon line by a nonlinear absorption method during magnetic scanning, *Opt. & Spectrosc. (USA)* **49**, 486–489.

Kowalski, J., Neumann, R., Suhr, H., Winkler, K., and zu Putlitz, G., 1978, Two-photon intracavity dye laser spectroscopy of the 4S and 3D term in $^{6,7}Li$, *Z. Phys. A* **287**, 247–253.

Kronfeldt, H.-D., Kropp, J.-R., and Winkler, R., 1982, The isotope shift of the a^{10}D-term of the configuration $4f^7 5d6s$ in Eu I, *Physica* **112C**, 138–144.

Kühl, T., Dabkiewicz, P., Duke, C., Fischer, H., Kluge, H.-J., Kremmling, H., and Otten, E.-W., 1977, Nuclear shape staggering in very neutron-deficient Hg isotopes detected by laser spectroscopy, *Phys. Rev. Lett.* **39**, 180–183.

Kuhn, H. G., 1969, *Atomic Spectra*, Longman's, London.

Kuhn, H. G., Baird, P. E. G., Brimicombe, M. W. S. M., Stacey, D. N., and Stacey, V., 1975, Evidence for alpha-particle structure in medium-heavy nuclei from optical isotope shifts, *Proc. R. Soc. London, Ser. A* **342**, 51–54.

Labarthe, J. J., 1973, Correlation effects on specific isotope shifts, *J. Phys. B* **6**, 1761–1770.

Labarthe, J. J., 1978, Calculation of hyperfine-structure second-order effects on the isotope shifts in Sm I, *J. Phys. B* **11**, L1–L4.

Lecordier, R., 1979, Determination des déplacements isotopique des isotopes impairs 123 et 125 du tellure pas rapport aux isotopes pairs 122, 124 et 126 ("odd–even staggering"), *Phys. Lett.* **72A**, 327–328.

Lecordier, R., and Helbert, J. M., 1978, Déplacement isotopique relatif dans les raies λ = 4048,9 Å, λ = 5449,8 Å et λ = 5479,1 Å du spectre II du tellure, *Physica* **94C**, 125–133.

Lee, P. L., and Boehm, F., 1973, X-ray isotope shifts and variations of nuclear charge radii in isotopes of Nd, Sm, Dy, Yb, and Pb, *Phys. Rev. C* **8**, 819–828.

Lee, P. L., Boehm, F., and Hahn, A. A., 1978, Variations of nuclear charge radii in mercury isotopes with $A = 198$, 199, 200, 201, 202, and 204 from x-ray isotope shifts, *Phys. Rev. C* **17**, 1859–1861.

Liberman, S., Pinard, J., Duong, H. T., Juncar, P., Pillet, P., Vialle, J.-L., Jacquinot, P., Touchard, F., Büttgenbach, S., Thibault, C., de Saint-Simon, M., Klapisch, R., Pesnelle, A., and Huber, G., 1980, Laser optical spectroscopy on francium D_2 resonance line, *Phys. Rev. A* **22**, 2732–2737.

Lindroth, E., and Mårtensson-Pendrill, A.-M., 1983, Calculation of the isotope shift in Na, *Z. Phys. A* **309**, 277–284.

Lorenzen, C.-J., and Niemax, K., 1982, Level isotope shifts of ^{6,7}Li, *J. Phys. B* **15**, L139–L145.

Lyman, E. M., Hanson, A. O., Scott, M. B., 1951, Scattering of 15.7-MeV electrons by nuclei, *Phys. Rev.* **84**, 626–634.

Lyons, D. R., Schawlow, A. L., and Yan, G.-Y., 1981, Doppler-free radiofrequency optogalvanic spectroscopy, *Opt. Commun.* **38**, 35–38.

Macagno, E. R., Bernow, S., Cheng, S. C., Devons, S., Duerdoth, I., Hitlin, D., Kast, J. W., Lee, W. Y., Rainwater, J., Wu, C. S., and Barrett, R. C., 1970, Muonic atoms. II Isotope shifts, *Phys. Rev. C* **1**, 1202–1221.

Magnante, P. C., and Stroke, H. H., 1969, Isotope shift between ^{209}Bi and 6.3-day ^{206}Bi, *J. Opt. Soc. Am.* **59**, 836–841.

Mariella, R., 1979, The isotope shift in the 2^2P states of lithium and spatially resolved laser-induced fluorescence, *Appl. Phys. Lett.* **35**, 580–582.

Mariño, C. A., Fülöp, G. F., Groner, W., Moskowitz, P. A., Redi, O., and Stroke, H. H., 1975, Nuclear magnetic moments of ^{205,207,209}Bi isotopes—hyperfine structure of the 15-day ^{205}Bi 3067-A line, *Phys. Rev. Lett.* **34**, 625–628.

Mårtensson, A.-M., 1979, An iterative, numeric procedure to obtain pair functions applied to two-electron systems, *J. Phys. B.* **12**, 3995–4012.

Mårtensson, A.-M., and Salomonson, S., 1982, Specific mass shifts in Li and K calculated using many-body perturbation theory, *J. Phys. B* **15**, 2115–2130.

Martin, W. C., Zalubas, R., and Hagan, L., 1978, *Atomic Energy Levels — The Rare-Earth Elements*, U.S. Government Printing Office, Washington, D.C.

Melissinos, A. C., and Davis, S. P., 1959, Dipole and quadrupole moments of the isomeric Hg^{197*} nucleus; isomeric isotope shift, *Phys. Rev.* **115**, 130–137.

Merton, T. R., 1915, On the spectra of ordinary lead and lead of radioactive origin, *Proc. R. Soc. London, Ser. A* **91**, 198–201.

Michelson, A. A., 1893, Comparison of the International Meter with the wave length of the light of cadmium, *Astron. Astrophys.* **12**, 556–560.

Miller, G. E., and Ross, J. S., 1976, Isotope shifts in the arc spectra of dysprosium, erbium, and ytterbium, *J. Opt. Soc. Am.* **66**, 585–589.

Milner, W. T., Bemis, C. E., Jr., and McGowan, F. K., 1977, Quadrupole and hexadacapole deformations in the actinide nuclei, *Phys. Rev. C* **16**, 1686–1687.

Moinester, M. A., Alster, J., Azuelos, G., Bellicard, J. B., Frois, B., Huet, M., Leconte, P., and Ho, P. X., 1981, Charge and transition densities for the samarium isotopes by electron scattering, *Phys. Rev. C* **24**, 80–88.

Moscatelli, F. A., Redi, O., Schönberger, P., Stroke, H. H., and Wiggins, R. L., 1982, Isotope shifts in the $6p^2\,^3P_0$–$6p7s\,^3P_1^0$ 283.3 nm line of natural lead, *J. Opt. Soc. Am.* **72**, 918–922.

Müller, G., and Winkler, R., 1975, Eine Hyperfeinstruktur-Analyse der Grundkonfigurationen $5d^86s^2$ und $5d^96s$ des Pt I Spektrums, *Z. Phys. A* **273**, 313–320.

Neijzen, J. H. M., and Dönszelmann, A., 1980, Hyperfine structure and isotope shift measurements in neutral gallium and indium with a pulsed dye laser, *Physica* **98C**, 235–241.

Neukammer, J., and Rinneberg, H., 1982, Hyperfine structure of perturbed $6sns\,^3S_1$ Rydberg states of barium, *J. Phys. B* **15**, L425–L429.

New, R., Griffith, J. A. R., Isaak, G. R., and Ralls, M. P., 1981, *J* dependence of the field and mass isotope shifts in the ground term of Sm I, *J. Phys. B* **14**, L135–L139.

Nicklas, J. P., and Treanor, C. E., 1958, Hartree–Fock functions and spectral isotope shifts for excited states of carbon and oxygen, *Phys. Rev.* **110**, 370–374.

Niemax, K., and Pendrill, L. R., 1980, Isotope shifts of individual nS and nD levels of atomic potassium, *J. Phys. B* **13**, L461–L465.

Nöldeke, G., Steudel, A., Wallach, K. E., and Walther, H., 1962, Die relative Isotopenlage des Os^{186}, *Z. Phys.* **170**, 22–25.

Odintsova, N. K., and Striganov, A. R., 1976, Isotope shift and deformation of gadolinium nuclei, *Opt. & Spectrosc. (USA)* **41**, 545–547.

Olin, A., Poffenberger, P. R., Beer, G. A., Macdonald, J. A., Mason, G. R., and Pearce, R. M., 1981, Measurement of pionic and muonic X rays in ^{10,11}B, *Nucl. Phys.* **A360**, 426–434.

Otten, E. W., 1981, Laser techniques in nuclear physics, *Nucl. Phys.* **A354**, 471c–496c.

Palmer, C. W. P., Baird, P. E. G., Nicol, J. L., Stacey, D. N., and Woodgate, G. K., 1982, Isotope shift in calcium by two-photon spectroscopy, *J. Phys. B* **15**, 993–995.

Palmer, C. W. P., and Stacey, D. N., 1982, Theory of anomalous isotope shifts in samarium, *J. Phys. B* **15**, 997–1005.

Pekeris, C. L., 1958, Ground state of two-electron atoms, *Phys. Rev.* **112**, 1649–1658.

Pekeris, C. L., 1959, 1^1S and 2^3S states of helium, *Phys. Rev.* **115**, 1216–1221.

Pekeris, C. L., 1962, Excited S states of helium, *Phys. Rev.* **127**, 509–511.

Pendrill, L. R., and Niemax, K., 1982, Isotope shifts of energy levels in ^{40}K and ^{39}K, *J. Phys. B* **15**, L147–L151.

Pfeiffer, H.-J., and Daniel, H., 1976, Elektromagnetische Kernradien der Stabilen Argon-Isotope, *Nucl. Phys.* **A264**, 498–506.

Pfeufer, V., Reimche, A., and Steudel, A., 1982, Crossed second-order effects in the isotope shift of Eu I, *Z. Phys. A* **308**, 185–186.

Powers, R. J., Barreau, P., Bihoreau, B., Miller, J., Morgenstern, J., Picard, J., and Roussel, L., 1979, A muonic x-ray study of the charge distribution of ^{144,148,150,152,154}Sm, *Nucl. Phys.* **A316**, 295–316.

Prasad, S. S., and Stewart, A. L., 1966, Isotope shift in Li and B^{2+}, *Proc. Phys. Soc.* **87**, 159–161.

Racah, G., 1932, Isotopic displacement and hyperfine structure, *Nature* **129**, 723–724.

Rajnak, K., and Fred, M., 1977, Correlation of isotope shifts with $|\psi(0)|^2$ for actinide configurations, *J. Opt. Soc. Am.* **67**, 1314–1323.

Rao, P. R., and Gluck, G., 1964, Isotope shifts in the Nd I spectrum, *Proc. R. Soc. London, Ser. A* **277**, 540–548.

Rinneberg, H., Neukammer, J., and Matthias, E., 1982, Isotope shifts of perturbed $6sns\ ^1S_0$ and $^3S_1$ Rydberg states of odd barium isotopes, *Z. Phys. A* **306**, 11–18.

Rosenthal, J. E., and Breit, G., 1932, The isotope shift in hyperfine structure, *Phys. Rev.* **41**, 459–470.

Ruckstuhl, W., Aas, B., Beer, W., Beltrami, I., de Boer, F. W. N., Bos, K., Goudsmit, P. F. A., Kiebele, U., Leisi, H. J., Strassner, G., Vacchi, A., and Weber, R., 1982, High-precision muonic x-ray measurement of the RMS charge radius of ^{12}C with a crystal spectrometer, *Phys. Rev. Lett.* **49**, 859–862.

Russell, A. S., and Rossi, R., 1912, An investigation of the spectrum of ionium, *Proc. R. Soc. London, Ser. A* **87**, 478–484.

Schaller, L. A., Dubler, T., Kaeser, K., Rinker, G. A., Jr., Robert-Tissot, B., Schellenberg, L., and Schneuwly, H., 1978, Nuclear charge radii from muonic x-ray transitions in F, Na, Al, Si, P, S and K, *Nucl. Phys.* **A300**, 225–234.

Schaller, L. A., Schellenberg, L., Phan, T. Q., Piller, G., Ruetschi, A., and Schneuwly, H., 1982, Nuclear charge radii of the carbon isotopes ^{12}C, ^{13}C and ^{14}C, *Nucl. Phys.* **A379**, 523–535.

Schaller, L. A., Schellenberg, L., Ruetschi, A., and Schneuwly, H., 1980, Nuclear charge radii from muonic X ray transitions in beryllium, boron, carbon and nitrogen, *Nucl. Phys.* **A343**, 333–346.

Schellenberg, L., Robert-Tissot, B., Käser, K., Schaller, L. A., Schneuwly, H., Fricke, G., Glückert, S. and Mallot, G., 1980, Systematics of nuclear charge radii of the stable molybdenum isotopes from muonic atoms, *Nucl. Phys.* **A333**, 333–342.

Schiff, B., Lifson, H., Pekeris, C. L., and Rabinowitz, P., 1965, $2^{1,3}P$, $3^{1,3}P$, and $4^{1,3}P$ states of He and the 2^1P state of Li^+, *Phys. Rev.* **140**, A1104–A1121.

Schroeder, D. J., and Mack, J. E., 1961, Isotope shift in the arc spectrum of nickel, *Phys. Rev.* **121**, 1726–1731.

Schüler, H., 1926, Über eine neue Lichtquelle und ihre Anwendungsmöglichkeiten, *Z. Phys.* **35**, 323–337.

Schüler, H., 1927, Weitere Untersuchungen am ersten Li-Funkenspectrum, *Z. Phys.* **42**, 487–494.

Schüler, H., and Jones, E. G., 1932, Hyperfeinstrukturen und Kernmomente des Quecksilbers, *Z. Phys.* **74**, 631–646.

Schüler, H., and Korsching, H., 1938, Magnetische Momente von ^{171,173}Yb und Isotopenverschiebung beim Yb I, *Z. Phys.* **111**, 386–390.

Segrè, E., 1977, *Nuclei and Particles*, The Benjamin/Cummings Publishing Company, Reading, Massachusetts.

Seltzer, E. C., 1969, *K* x-ray isotope shifts, *Phys. Rev.* **188**, 1916–1919.

Shafer, J. H., 1971, Optical heterodyne measurement of xenon isotope shifts, *Phys. Rev. A* **3**, 752–757.

Shenoy, G. K., and Wagner, F. E., eds., 1978, *Mössbauer Isomer Shifts*, North Holland, Amsterdam.

Shera, E. B., Ritter, E. T., Perkins, R. B., Rinker, G. A., Wagner, L. K., Wohlfahrt, H. D., Fricke, G., and Steffen, R. M., 1976, Systematics of nuclear charge distributions in Fe, Co, Ni, Cu, and Zn deduced from muonic x-ray measurements, *Phys. Rev. C* **14**, 731–747.

Shera, E. B., Wohlfahrt, H. D., Hoehn, M. V., and Tanaka, Y., 1982, Charge distributions of barium isotopes from muonic X rays. *Phys. Lett.* **112B**, 124–128.

Sick, I., 1982, Precise nuclear radii from electron scattering, *Phys. Lett.* **116B**, 212–214.

Siegel, A., Lawler, J. E., Couillard, B., and Hansch, T. W., 1981, Doppler-free spectroscopy in a hollow-cathode discharge: Isotope-shift measurements in molybdenum, *Phys. Rev. A* **23**, 2457–2461.

Silver, J. D., and Stacey, D. N., 1973a, Isotope shift and hyperfine structure in the atomic spectrum of tin, *Proc. R. Soc. London, Ser. A* **332**, 129–138.

Silver, J. D., and Stacey, D. N., 1973b, Isotope effects in the nuclear charge distribution in tin, *Proc. R. Soc. London, Ser. A* **332**, 139–150.

Soddy, F., 1913, Intra-atomic charge, *Nature* **92**, 399–400.

Stacey, D. N., 1966, Isotope shifts and nuclear charge distributions, *Rep. Prog. Phys.* **29**, 171–215.

Stacey, D. N., 1971, Isotope shifts in electronic and muonic atoms, with application to neodymium, *J. Phys. B* **4**, 969–975.

Steudel, A., Triebe, U., and Wendlandt, D., 1980, Isotope shift in Ni I and changes in mean-square nuclear charge radii of the stable Ni isotopes, *Z. Phys. A* **296**, 189–193.

Stone, A. P., 1959, Calculations of isotope shifts in the spectrum of helium, *Proc. Phys. Soc.* **68**, 1152–1156.

Stone, A. P., 1961, Nuclear and relativistic effects in atomic spectra, *Proc. Phys. Soc.* **77**, 786–796.

Stone, A. P., 1963, Nuclear and relativistic effects in atomic spectra: II, *Proc. Phys. Soc.* **81**, 868–876.

Striganov, A. R., Katulin, V. A., and Eliseev, V. V., 1962, Peculiarities of the isotopic shift in the samarium spectrum, *Opt. & Spectrosc. (USA)* **12**, 91–94.

Stroke, H. H., Proetel, D., and Kluge, H.-J., 1979, Odd–even staggering in mercury isotope shifts: evidence for Coriolis effects in particle-core coupling, *Phys. Lett.* **82B**, 204–207.

Thibault, C., Touchard, F., Büttgenbach, S., Klapisch, R., de Saint Simon, M., Duong, H. T.,

Jacquinot, P., Juncar, P., Liberman, S., Pillet, P., Pinard, J., Vialle, J. L., Pesnelle, A. and Huber, G., 1981a, Hyperfine structure and isotope shift of the D_2 line of $^{76-98}$Rb and some of their isomers, *Phys. Rev. C* **23**, 2720–2729.

Thibault, C., Touchard, F., Büttgenbach, S., Klapisch, R., de Saint Simon, M., Duong, H. T., Jacquinot, P., Juncar, P., Liberman, S., Pillet, P., Pinard, J., Vialle, J. L., Pesnelle, A., the ISOLDE Collaboration, and Huber, G., 1981b, Hyperfine structure and isotope shift of the D_2 line of $^{118-145}$Cs and some of their isomers, *Nucl. Phys.* **A367**, 1–12.

Thompson, R. C., Anselment, M., Bekk, K., Göring, S., Hanser, A., Meisel, G., Rebel, H., Schatz, G., and Brown, B. A., 1983, High-resolution measurements of isotope shifts and hyperfine structure in stable and radioactive lead isotopes, *J. Phys. G* **9**, 443–458.

Tomkins, F., and Gerstenkorn, S., 1967, Déplacement isotopique relatif dans le spectre d'arc du plutonium. Effet spécifique, *C. R. Hebd. Seances Acad. Sci. Ser. B* **265**, 1311–1313.

Tomlinson, W. J., III, and Stroke, H. H., 1962, Nuclear isomer shift in the optical spectrum of Hg^{195}: interpretation of the odd–even staggering effect in isotope shift, *Phys. Rev. Lett.* **8**, 436–438.

Touchard, F., Guimbal, P., Büttgenbach, S., Klapisch, R., de Saint Simon, M., Serre, J. M., Thibault, C., Duong, H. T., Juncar, P., Liberman, S., Pinard, J., and Vialle, J. L., 1982a, Isotope shifts and hyperfine structure of $^{38-47}$K by laser spectroscopy, *Phys. Lett.* **108B**, 169–171.

Touchard, F., Serre, J. M., Büttgenbach, S., Guimbal, P., Klapisch, R., de Saint Simon, M., Thibault, C., Duong, H. T., Juncar, P., Liberman, S., Pinard, J., and Vialle, J. L., 1982b, Electric quadrupole moments and isotope shifts of radioactive sodium isotopes, *Phys. Rev. C* **25**, 2756–2770.

Träger, F., 1981, On the charge distribution of calcium nuclei, *Z. Phys. A* **299**, 33–39.

Trees, R. E., and Harvey, M. M., 1952, Low even configurations of the first spectrum of molybdenum (Mo I), *J. Res. Nat. Bur. Std.* **49**, 397–408.

Urey, H. C. , Brickwedde, F. G., and Murphy, G. M., 1932, A hydrogen isotope of mass 2 and its concentration, *Phys. Rev.* **40**, 1–15.

van Eijk, C. W. E., and Schutte, F., 1970, The $K\alpha_1$ X ray isotopic shift for ^{162}Dy–^{164}Dy and the differences in nuclear charge radii of Dy isotopes, *Nucl. Phys.* **A151**, 459–464.

van Eijk, C. W. E., and Visscher, M. J. C., 1970, The $K\alpha_1$ X ray isotopic shift for ^{140}Ce–^{142}Ce and the differences in nuclear charge radii of Ce isotopes, *Phys. Lett.* **34B**, 349–350.

van Eijk, C. W. E., Wijnhorst, J., Popelier, M. A., and Gillespie, W. A., 1979, The differences in nuclear charge radius between ^{110,112,114,116}Cd, *J. Phys. G* **5**, 315–317.

Van Hove, M., Borghs, G., De Bisschop, P., and Silverans, R. E., 1982, J-dependent isotope shifts in the 5d ^{2}D$_J$ doublet of barium II, *J. Phys. B* **15**, 1805–1809.

van Leeuwen, K. A. H., Eliel, E. R., Post, B. H., and Hogervorst, W., 1981, High-resolution measurements of hyperfine structure and isotope shifts in 9 spectral lines of Nd I, *Z. Phys. A* **301**, 95–99.

Vinti, J. P., 1939, Isotope shift in magnesium, *Phys. Rev.* **56**, 1120–1132.

Vizbaraite, J. J., Eriksonas, K. M., and Boruta, J. J., 1979, A procedure recommended for theoretical research on isotope shift in complex electronic configurations, *Sov. Phys. Collect (USA)* **19**, Part 4, 14–19.

Vogel, P., 1974, Tables of electron screening and higher-order vacuum polarization potentials in mesic atoms, *At. Data & Nucl. Data Tables* **14**, 599–604.

Wagner, S., 1955, Zur Isotopieverschiebung im Cu I-Spektrum, *Z. Phys.* **141**, 122–145.

Wagstaff, C. E., and Dunn, M. H., 1980, Saturated absorption spectroscopy in the UV with a continuous-wave frequency doubled ring dye laser, even isotope shifts in Hg I, *Opt. Commun.* **35**, 353–358.

Wenz, R., Matthias, E., Rinneberg, H., and Schneider, F., 1980, Further evidence for a linear relationship between changes in nuclear charge radii and binding energies per nucleon, *Z. Phys. A* **295**, 303–304.

Wenz, R., Timmermann, A., and Matthias, E., 1981, Subshell effect in mean square charge radii of stable even cadmium isotopes, *Z. Phys. A* **303**, 87–95.

Wieman, C., and Hänsch, T. W., 1980, Precision measurements of the 1S Lamb shift and of the 1S–2S isotope shift of hydrogen and deuterium, *Phys. Rev. A* **22**, 192–205.

Wilson, M., 1968, *Ab initio* calculation of screening effects on $|\psi(0)|^2$ for heavy atoms, *Phys. Rev.* **176**, 58–63.

Wilson, M., 1972, *Ab initio* calculation of core relaxation and screening effects on $|\psi(0)|^2$ for Sm and Eu, *J. Phys. B* **5**, 218–228.

Wilson, M., 1978a, *Ab initio* calculation of isotope shifts in Ba II, *Phys. Lett.* **65A**, 213–214.

Wilson, M., 1978b, *Ab initio* calculation of isotope shifts in Ce II, *Physica* **95C**, 129–133.

Wilson, M., 1982, Pseudo-relativistic screening ratios for Eu, *J. Phys. B* **15**, L375–L377.

Wohlfahrt, H. D., Schwentker, O., Fricke, G., Andresen, H. G., and Shera, E. B., 1980, Systematics of nuclear charge distributions in the mass 60 region from elastic electron scattering and muonic x-ray measurements, *Phys. Rev. C* **22**, 264–283.

Wohlfahrt, H. D., Shera, E. B., Hoehn, M. V., Yamazaki, Y. and Steffen, R. M., 1981, Nuclear charge distributions in $1f_{7/2}$-shell nuclei from muonic x-ray measurements, *Phys. Rev. C* **23**, 533–548.

Worden, E. F., and Conway, J. G., 1976, Energy levels of the first spectrum of curium, Cm I, *J. Opt. Soc. Am.* **66**, 109–121.

Wu, C. S., and Wilets, L., 1969, Muonic atoms and nuclear structure, *Ann. Rev. Nucl. Sci.* **19**, 527–606.

Yamazaki, Y., Shera, E. B., Hoehn, M. V., and Steffen, R. M., 1978, Measurement and model-independent analysis of the X rays of muonic ^{150}Sm and ^{152}Sm, *Phys. Rev. C* **18**, 1474–1496.

Zaal, G. J., Hogervorst, W., Eliel, E. R., van Leeuwen, K. A. H., and Blok, J., 1979, A high-resolution study of the transitions $4f^7 6s^2 \rightarrow 4f^7 6s6p$ in the Eu I Spectrum, *Z. Phys. A* **290**, 339–344.

Zaal, G. J., Hogervorst, W., Eliel, E. R., van Leeuwen, K. A. H., and Blok, J., 1980, A study of the spectrum of natural dysprosium with the laser-atomic-beam technique. I. Isotope shifts, *J. Phys. B* **13**, 2185–2194.

Zalubas, R., and Albright, A., 1980, *Bibliography on Atomic Energy Levels and Spectra: July 1975 through June 1979*, U.S. Government Printing Office, Washington, D.C.

Zehnder, A., Boehm, F., Dey, W., Engfer, R., Walter, H. K., and Vuilleumier, J. L., 1975, Charge parameters, isotope shifts, quadrupole moments, and nuclear excitation in muonic $^{170-174,176}$Yb, *Nucl. Phys.* **A254**, 315–340.

Zimmermann, D., Zimmermann, P., Aepfelbach, G. and Kuhnert, A., 1980, Isotope shift and hyperfine structure of the transition $5d6s^2\, ^2D_{3/2}$–$5d6s6p^4 F_{3/2}$ of Lu^{175} and Lu^{176}, *Z. Phys. A* **295**, 307–310.

AUTHOR INDEX

References are given only for those cases where the author is not given by name, but only included in "*et al.*"

Aas, B. (Ruckstuhl *et al.*, 1982), 100
Accad, Y., 22–24
Aepfelbach, G. (Zimmermann *et al.*, 1980), 143
Ahmad, S. A., 136, 140, 162
Albright, A., 177
Alster, J. (Moinester *et al.*, 1981), 136
Alvarez, E., 125
Amin, S. R., 97
Anderson, H. L. (Kessler *et al.*, 1975), 75, 76, 154
Andl, A. (Bekk *et al.*, 1979), 11, 67, 70, 71, 77, 87, 107, 127, 166
Andresen, H. G. (Wohlfahrt *et al.*, 1980), 59, 61, 75, 110, 112, 167
Anigstein, R., 78
Anselment, M. (Thompson *et al.*, 1983), 57, 77, 89, 150, 154
Arnesen, A. (Alvarez *et al.*, 1979), 125
Aronberg, L., 6
Aufmuth, P., 52, 53, 117–119, 141, 162
Aumann, D. C. (Fischer *et al.*, 1975a), 133, 134
Azuelos, G. (Moinester *et al.*, 1981), 136

Babushkin, F. A., 8, 36, 41, 42
Backenstoss, G., 100
Baird, P. E. G. (Kuhn *et al.*, 1975) (Palmer *et al.*, 1982), 11, 67, 70, 71, 87, 91, 107, 108, 120, 127, 146, 168
Ballik, E. A., 101
Barbier, L., 89
Bardin, T. T., 155
Barreau, P. (Powers *et al.*, 1979), 76, 79, 94, 136, 137
Barrett, R. C. (Bardin *et al.*, 1967) (Hitlin *et al.*, 1970) (Macagno *et al.*, 1970), 11, 35, 59–61, 74–76, 78, 81, 118, 121, 122, 135, 143, 144, 155, 177

Bauche, J. (Biraben *et al.*, 1976) (de Clercq *et al.*, 1981) (Eichhorn *et al.*, 1982) (Gluck *et al.*, 1964) (Grethen *et al.*, 1980) (Jackson *et al.*, 1975), 9–10, 13, 22, 23, 26, 27, 31, 32, 34, 44, 46, 52, 65, 88, 91, 99–105, 107, 108, 110–112, 116, 122, 124, 136, 139, 145, 150, 158, 162, 177
Bayer, R., 25, 98
Beene, J. R. (Bemis *et al.*, 1979), 89, 159, 160
Beer, G. A. (Olin *et al.*, 1981), 99
Beer, W. (Ruckstuhl *et al.*, 1982), 100
Beigang, R., 107, 108, 116
Bekk, K. (Andl *et al.*, 1982) (Thompson *et al.*, 1983), 11, 57, 67, 70, 71, 77, 87, 89, 107, 127, 150, 154, 166
Bellicard, J. B. (Cavedon *et al.*, 1982) (Moinester *et al.*, 1981), 136, 168
Beltrami, I. (Ruckstuhl *et al.*, 1982), 100
Bemis, C. E., Jr. (Milner *et al.*, 1977), 82, 89, 159, 160
Bengston, A. (Alvarez *et al.*, 1979), 125
Bergmann, E., 107
Bergstrom, J. C. (Briscoe *et al.*, 1980), 104
Bernow, S. (Hitlin *et al.*, 1970) (Kast *et al.*, 1971) (Macagno *et al.*, 1970), 78, 81, 118, 121, 122, 135, 144
Bethe, H. A., 15, 20
Bhattacharjee, S. K., 93, 121
Bihoreau, B. (Powers *et al.*, 1979), 76, 79, 136, 137
Biraben, F. (de Clercq *et al.*, 1981) (Giacobino *et al.*, 1979), 22, 23, 31, 34, 91
Bishop, D. C., 83, 122
Blaise, J., 85, 156–159
Blok, J., (Zaal *et al.*, 1979; 1980), 88, 139–141
Bloomfield, L. A. (Coillaud *et al.*, 1980), 90, 148

ELEMENT INDEX

SUBJECT INDEX

Alkali-metal spectra (table of isotope shifts), 30
Alpha-particle structures, 168
Analysis of isotope shift data, 127–131
Analysis of spectra, 162
Anomalous isotope shifts of mixed levels, 137–139
Atomic beam, 7, 86
Atomic unit of energy, 32, 33

Barrett moment, 59, 60
β, the screening factor, 47
β_2, the nuclear quadrupole deformation parameter, 50
Bohr shift, 15
Brix and Kopfermann diagram, 163, 164
Bubble nuclei, 73, 74

$c - t$ plot, 75
Charge density at nucleus
  of $p_{1/2}$ electrons, 53, 54
  of s electrons, 38, 42–49
Collinear laser spectroscopy, 88, 104
Coulomb excitation, 52, 80
Crossed second-order effect, 140

Deformed nuclei, 7, 50–52
Diffraction grating, 85
Doppler-free two-photon spectroscopy, 91, 92
Doppler width, 83, 84, 86

Elastic electron scattering, 73, 74
Electron screening, 6, 9, 46–49, 162
Electron screening factor (β), 9, 10, 46–49
Electron screening ratios, 47–49
Electron-to-proton mass ratio, 16, 17, 161, 175
Electronic factor (F), 42, 57
Enrichment of isotopes, 140, 141
Equivalent uniform radius of nucleus, 36
Errors, 3
Exchange coupling between electrons, 20

Fast atomic beams, 88, 104
Fermi nuclear charge distribution, 36, 59, 60
Field ionization, 89
Field shift (FS), 2
Field shift functions [including $f(Z)$], 42, 43
Fourier transform spectroscopy, 10, 85
Frequency doubling, 89

Generalized moment of the nuclear charge distribution, 59, 60
Goudsmit–Fermi–Segrè formula, 43–45

HFR pseudorelativistic method, 49
Hartree–Fock values of electron charge density, 47, 48
Hollow cathode lamp, 6, 83

Interacting boson model, 172
Intermodulated fluorescence spectroscopy, 92, 119
Intrinsic quadrupole moment, 51, 52
Isomer shift, 76–80, 148, 149
Isotope shift constant (C), 41
Isotope shift discrepancy, 166

J-dependent field shifts, 132, 136, 139–141, 145
J-dependent mass shifts, 31, 34, 139

k factor (specific mass shift), 32–34
Kayser unit of wavenumber (K), 4, 176
King plot, 64–66

Laser-induced atomic beam fluorescence spectroscopy, 87

Magic number of nucleons, 165
Magnetic hyperfine structure (hfs), 80, 81
Magnetic rotation spectroscopy, 113
Mass polarization correction, 22
Mass shift (MS), 1